Mobile Design and Development

Mobile Design and Development

Brian Fling

Beijing · Cambridge · Farnham · Köln · Sebastopol · Taipei · Tokyo

Mobile Design and Development
by Brian Fling

Copyright © 2009 Brian Fling. All rights reserved.
Printed in the United States of America.

Published by O'Reilly Media, Inc., 1005 Gravenstein Highway North, Sebastopol, CA 95472.

O'Reilly books may be purchased for educational, business, or sales promotional use. Online editions are also available for most titles (*http://my.safaribooksonline.com*). For more information, contact our corporate/institutional sales department: (800) 998-9938 or *corporate@oreilly.com*.

Editor: Steven Weiss	**Indexer:** John Bickelhaupt
Production Editor: Loranah Dimant	**Cover Designer:** Karen Montgomery
Copyeditor: Nancy Kotary	**Interior Designer:** David Futato
Proofreader: Sada Preisch	**Illustrator:** Robert Romano

Printing History:

August 2009: First Edition.

RepKover™

This book uses RepKover™, a durable and flexible lay-flat binding.

ISBN: 978-0-596-15544-5

[M]

1280719380

[8/10]

Table of Contents

Preface xiii

1. **A Brief History of Mobile** 1
   In the Beginning   1
   The Evolution of Devices   3
       The Brick Era   4
       The Candy Bar Era   5
       The Feature Phone Era   6
       The Smartphone Era   8
       The Touch Era   10

2. **The Mobile Ecosystem** 13
   Operators   14
   Networks   17
   Devices   18
   Platforms   20
       Licensed   20
       Proprietary   21
       Open Source   21
   Operating Systems   22
   Application Frameworks   22
       Java   23
       S60   23
       BREW   23
       Flash Lite   23
       Windows Mobile   24
       Cocoa Touch   24
       Android SDK   24
       Web Runtimes (WRTs)   24
       WebKit   24
       The Web   25

Applications 25
Services 26

3. Why Mobile? **29**
Size and Scope of the Mobile Market 30
The Addressable Mobile Market 31
   High-End Versus Low-End Devices 32
   Best-selling Versus Free 34
   Mobile Web Versus Native Applications 34
   Touch Versus D-Pad 34
Mobile As a Medium 34
   The Printing Press 35
   Recordings 35
   Cinema 35
   Radio 36
   Television 36
   The Internet 36
   Mobile 37
   Mobile's Unique Benefits 39
The Eighth Mass Medium: What's Next? 40
Ubiquity Starts with the Mobile Web 41

4. Designing for Context **45**
Thinking in Context 46
   Context with a Capital C 47
   Context with a Lowercase c 52
Taking the Next Steps 55

5. Developing a Mobile Strategy **57**
New Rules 59
   Rule #1: Forget What You Think You Know 59
   Rule #2: Believe What You See, Not What You Read 60
   Rule #3: Constraints Never Come First 61
   Rule #4: Focus on Context, Goals, and Needs 63
   Rule #5: You Can't Support Everything 65
   Rule #6: Don't Convert, Create 66
   Rule #7: Keep It Simple 67
Summary 67

6. Types of Mobile Applications **69**
Mobile Application Medium Types 70
   SMS 70
   Mobile Websites 71

Mobile Web Widgets 73
Mobile Web Applications 75
Native Applications 77
Games 79
Mobile Application Media Matrix 80
Application Context 81
Utility Context 81
Locale Context 83
Informative Applications 84
Productivity Application Context 85
Immersive Full-Screen Applications 87
Application Context Matrix 88

7. **Mobile Information Architecture** **89**
What Is Information Architecture? 89
Mobile Information Architecture 91
Keeping It Simple 93
Site Maps 94
Clickstreams 98
Wireframes 101
Prototyping 103
Different Information Architecture for Different Devices 105
The Design Myth 106

8. **Mobile Design** **109**
Interpreting Design 111
The Mobile Design Tent-Pole 112
Designing for the Best Possible Experience 115
The Elements of Mobile Design 116
Context 116
Message 117
Look and Feel 118
Layout 121
Color 125
Typography 129
Graphics 134
Mobile Design Tools 137
Designing for the Right Device 138
Designing for Different Screen Sizes 139

9. **Mobile Web Apps Versus Native Applications** **143**
The Ubiquity Principle 143
Fragmentation 144

The Web 144
Control 144
Consumer Expectations 145
Ubiquity in the Mobile Web 145
When to Make a Native Application 146
Charging for It 146
Creating a Game 147
Using Specific Locations 147
Using Cameras 147
Using Accelerometers 148
Accessing the Filesystems 148
Offline Users 149
When to Make a Mobile Web Application 150

10. Mobile 2.0 .... **153**
What Is Mobile 2.0? 154
Mobile 2.0: The Convergence of the Web and Mobile 155
The Mobile Web Browser As the Next Killer App 155
Mobile Web Applications Are the Future 156
JavaScript Is the Next Frontier 157
The Mobile User Experience Is Awful 158
Mobile Widgets Are the Next Big Thing 158
Carrier Is the New "C" Word 159
Mobile Needs to Check Its Ego 159
We Are Creators, Not Consumers 160

11. Mobile Web Development **163**
Web Standards 164
Designing for Multiple Mobile Browsers 165
Progressive Enhancement 165
DIAL 167
Designing for Multiple Displays 168
Device Plans 169
The Device Matrix 170
Markup 172
XHTML-MP Overview 172
Document Structure 173
Text Elements 176
Creating Links 179
Images and Objects 180
Tables 182
Frames 183
Forms 183

Other Recommendations 184
CSS: Cascading Style Sheets 185
    Wireless CSS and CSS-MP 186
    Box Model 186
    Selectors 187
    Font and Text Properties 189
    Basic Box Properties 192
    Color and Backgrounds 194
    Positioning and Page Flow 194
JavaScript 196

12. iPhone Web Apps **199**
Why WebKit? 200
    A Brief History of WebKit 200
    Background As a Mobile Browser 201
What Makes It a Mobile Web App? 204
    The Page Model 205
Markup 206
    XHTML 206
    HTML5 209
CSS 213
    CSS2 214
    CSS3 216
    Visual Effects 221
JavaScript 225
    DHTML 226
    Ajax 226
    Multitouch 227
    Fixed Footer 227
Creating a Mobile Web App 228
    Defining the Viewport 229
    Full-Screen Mode 230
    Changing the Status Bar Appearance 230
    Adding an Icon 231
Web Apps As Native Apps 231
PhoneGap 232
Tools and Libraries 233
    iPhone GUI PSD 233
    iUI 234
    jQTouch 234

13. Adapting to Devices **237**
Why Is Adaptation a "Necessity"? 240

Strategy #1: Do Nothing   242
    Five Assumptions About One Web   242
    The One Web Aftermath   243
    Using This Strategy with Media Queries   244
Strategy #2: Progressive Enhancement   244
    The Handheld Media Type   245
    Layering Multiple Stylesheets for Multiple Devices   246
Strategy #3: Device Targeting   247
    The Device Detection Dilemma   248
    Andy Moore's Mobile Browser Detection   248
    Greg Mulmash's Mobile Browser Detection   249
    The Switcher   250
    htaccess-Based Device Detection   250
    JavaScript-Based Device Detection   251
    Reverse Device Detection   251
    WordPress Mobile Plugin   252
    dotMobi WordPress Mobile Pack   252
    Mobile Fu   253
    And Many More...   253
Strategy #4: Full Adaptation   253
    Working "On Deck"   254
    Working "Off Deck"   255
    WURFL   255
    DeviceAtlas   256
    Volantis   258
    WALL and WNG   258
    Yahoo! Blueprint   259
    Netbiscuits   259
    MobileAware   260
    Mobify   260
What Domain Do I Use?   261
    m.domain.com   262
    domain.com/mobile or domain.com/m   262
    domain.mobi   263
Taking the Next Step   263

14. Making Money in Mobile **265**
Working with Operators   268
    The Deck   268
    ARPU   270
    BoBo   271
Working with an App Store   271
    What About the Mobile Web?   273

Add Advertising   273
   AdMob and Google AdSense   274
   The Mobile Marketing Association   274
Invent a New Model   275

15. Supporting Devices **277**
   Having a Device Plan   278
      Deciding What to Support   278
      Example Device Plans   279
   Device Testing   282
      Access to Devices   282
      Estimating the Testing Effort   284
      Creating a Test Plan   285
      Creating a Test Portal   287
   Desktop Testing   288
      Frames   288
      Opera   288
      WebKit   289
      Firefox   291
      Collecting User Agents   292
      Simulators and Emulators   292
      Remote Access   295
   Usability Testing   295
      Mobile Usability Test Tips and Tricks   297

16. The Future of Mobile **299**
   The Opportunity for Change   300

Index **303**

Preface

I'll be honest: I'm an introduction-skipper. When I sit down with a technical book, I skip right past the introduction or preface and go straight for the goods. If it doesn't begin with the words "Chapter," then I figure I can probably move on and not miss anything crucial. This is not, however, one of those books.

Mobile design and development is about context, so it is somewhat fitting that the introduction of a book by the same name would establish context for the pages to follow. Before you dive into the wonderful world of mobile, I want to stress the scope of the medium and therefore of this book.

People don't seem to realize that mobile as a whole is really, really big. When someone says the word "mobile," they could be referring to devices, networks, services, the mobile web—even native applications like iPhone apps or a dozen other parts of a vast ecosystem. It isn't unlike saying that all the various technologies required to create a simple web page can simply be referred to as "the Web." There is obviously a lot more to it, but at the end of the day we just want it to work.

If there is one thing I've learned from my adventures in the mobile industry over the last decade, it is that in order to understand mobile, in order to make it work for you and for your users, you need understand three basic principles of mobile:

You need to know the different facets of the mobile medium
> There are many variables that can change the overall experience for the user, both good and bad. Understanding what they are, and how they might affect your project, at least at a basic level, can prevent serious and expensive problems later on

You need to know how to leverage mobile technologies to address context
> Context is the mental model in which information is understood. It is the key concept that makes mobile such a powerful and useful medium for millions of people around the world. But addressing context takes not just an understanding of user-centered design principles, but of what roles mobile devices play in people's lives.

You need to know how to leverage the right mobile technology for the need
> Here comes the tricky part: although there are numerous technologies within the mobile ecosystem that work well to address particular problems, finding the right

one for your users, your business, or your development resources can be incredibly hard. If you don't understand the pros and cons of each, it can be difficult to make the right investment, be it time or money, at the right time.

The first half of this book will cover these first two principles; the second half will focus on the last principle. When I sat down to write this book, I knew it would be impossible to cover every technology of mobile in detail. Therefore I'll focus a lot of this book on the mobile web as the only ubiquitous platform across all mobile devices around the world. Not only does the majority of the mobile community believe that the mobile web is the future of the mobile medium, but I've also found it to have the highest return on investment, be it in terms of money, user satisfaction, or development time.

I wrote this book to be a beginning—your beginning in mobile—and to give you all the information you need to know in order to start thinking of your site, application, or business in the mobile context. At the end, you should have a firm understanding of how mobile works and how to start designing and developing for it.

Who This Book Is For

I wrote this book to have something for everyone interested in designing in developing for mobile devices, regardless of experience and regardless of the application. The first half is a crash course in the mobile ecosystem: how to develop a strategy, address the mobile context—even how to decide which of the multiple mobile application types is best for you, and finally, how to create a user experience for it. The second half is focused on using these principles to make a mobile website or web app.

How This Book Is Organized

The chapters in this book are organized as follows:

Chapter 1, *A Brief History of Mobile*
> In this chapter, I'll provide a quick introduction into how mobile devices have evolved from phones to the pocket-sized computers of today and a look into where they are headed.

Chapter 2, *The Mobile Ecosystem*
> In this chapter, I'll give you a summary of the multiple layers of the mobile ecosystem and the role that each of them plays in getting your work into the hands and handsets of your users.

Chapter 3, *Why Mobile?*
> This chapter explores the importance of mobile around the world. I'll compare mobile to traditional media like print, television, and the Web, and explain some of the unique benefits of the mobile medium.

Chapter 4, *Designing for Context*

Creating mobile experience starts with addressing the context of the user. In this chapter, I'll discuss the different types of context and include some interesting examples of how mobile devices can address it.

Chapter 5, *Developing a Mobile Strategy*

In this chapter, I'll discuss how to create a user-centered, context-based mobile strategy. I'll include a few rules to help you make sure that your project starts off right and has the steam to get where it needs to go.

Chapter 6, *Types of Mobile Applications*

Many mobile projects fail because they aren't designed for the right type of application context. In this chapter, I'll explain the different types of applications, their pros and cons, and how to decide which is right for you.

Chapter 7, *Mobile Information Architecture*

In this chapter, I'll discuss how to structure the information in your product for the mobile context by using various deliverables to define your mobile experience.

Chapter 8, *Mobile Design*

In this chapter, I'll discuss how to create the best possible mobile experience and discuss the principles and techniques of how to create a design for mobile devices.

Chapter 9, *Mobile Web Apps Versus Native Applications*

In this chapter, I'll compare mobile web applications to applications written specifically for a particular mobile platform, including many of the pros and cons.

Chapter 10, *Mobile 2.0*

In this chapter, I'll discuss the concept of Mobile 2.0 and the importance of the mobile web to the future of mobile, as well as some of its challenges.

Chapter 11, *Mobile Web Development*

In this chapter, I will cover the mobile standards for various devices and provide an explanation of each of them in detail as well as how they are supported across multiple devices.

Chapter 12, *iPhone Web Apps*

In many ways, the iPhone is changing mobile for the better, leading innovation in the future of the mobile web. In this chapter, I will talk specifically about how to create iPhone web applications and how to make them work on other popular devices as well.

Chapter 13, *Adapting to Devices*

One of the greatest challenges in mobile is adapting to multiple devices. In this chapter, I will discuss some of the common techniques and a few services that can help.

Chapter 14, *Making Money in Mobile*

Once you have your product, it is time to publish it and, in most cases, try to make money from it. In this chapter, I will discuss the options and some of the common pitfalls to avoid.

Chapter 15, *Supporting Devices*
> In this chapter, I will discuss how to test and support multiple mobile devices, including a few tips and tricks of the trade.

Chapter 16, *The Future of Mobile*
> Finally, I will provide some of my thoughts as to the future of mobile and the next evolution of the Web.

Conventions Used in This Book

The following typographical conventions are used in this book:

Italic
> Indicates new terms, URLs, email addresses, filenames, and file extensions.

`Constant width`
> Used for program listings, as well as within paragraphs to refer to program elements such as variable or function names, databases, data types, environment variables, statements, and keywords.

`Constant width bold`
> Shows commands or other text that should be typed literally by the user.

`Constant width italic`
> Shows text that should be replaced with user-supplied values or by values determined by context.

 This icon signifies a tip, suggestion, or general note.

Using Code Examples

This book is here to help you get your job done. In general, you may use the code in this book in your programs and documentation. You do not need to contact us for permission unless you're reproducing a significant portion of the code. For example, writing a program that uses several chunks of code from this book does not require permission. Selling or distributing a CD-ROM of examples from O'Reilly books does require permission. Answering a question by citing this book and quoting example code does not require permission. Incorporating a significant amount of example code from this book into your product's documentation does require permission.

We appreciate, but do not require, attribution. An attribution usually includes the title, author, publisher, and ISBN. For example: "*Mobile Design and Development*, by Brian Fling. Copyright 2009 Brian Fling, 978-0-596-15544-5."

If you feel your use of code examples falls outside fair use or the permission given here, feel free to contact us at *permissions@oreilly.com*.

How to Contact Us

Please address comments and questions concerning this book to the publisher:

O'Reilly Media, Inc.
1005 Gravenstein Highway North
Sebastopol, CA 95472
800-998-9938 (in the United States or Canada)
707-829-0515 (international or local)
707-829-0104 (fax)

We have a web page for this book, where we list errata, examples, and any additional information. You can access this page at:

http://oreilly.com/catalog/9780596155445

To comment or ask technical questions about this book, send email to:

bookquestions@oreilly.com

For more information about our books, conferences, Resource Centers, and the O'Reilly Network, see our website at:

http://oreilly.com

Safari® Books Online

Safari Safari Books Online is an on-demand digital library that lets you easily search over 7,500 technology and creative reference books and videos to find the answers you need quickly.

With a subscription, you can read any page and watch any video from our library online. Read books on your cell phone and mobile devices. Access new titles before they are available for print, and get exclusive access to manuscripts in development and post feedback for the authors. Copy and paste code samples, organize your favorites, download chapters, bookmark key sections, create notes, print out pages, and benefit from tons of other time-saving features.

O'Reilly Media has uploaded this book to the Safari Books Online service. To have full digital access to this book and others on similar topics from O'Reilly and other publishers, sign up for free at *http://my.safaribooksonline.com*.

Acknowledgments

It amazes me the number of people required to get a book like this into your hands. To say that a book would not be possible without the following people feels like the understatement of a lifetime. I would like to thank all the people who sent me kind words of support throughout the creation of the book. I especially want to call out a few people in particular to whom I owe an enormous debt and my eternal gratitude.

Thank you to my friends in the mobile community who offered support and advice throughout the creation of this book: Kelly Goto, Barbara Ballard, Bryan Reiger, Chris Mills, Ronan Cremin, Mike Rowehl, Tony Fish, Dan Saffer, Matt May, Rudy De Waele, Katrin Verclas. And to my "nemesis" David Storey at Opera: thanks for helping to keep me honest.

A special thank you to all the people at my company, pinch/zoom, who had to work so hard to keep our little company going while I was "away": Garrett Murray, Charlie Barr, Cheryl Gledhill, Scott Gledhill, Jim Dovey, David Kaneda, and Tim Connor.

This book would not have been possible without the advice, support, and guidance of the technical editors and contributors:

- Daniel Appelquist: a guiding light in the future of the mobile web. If we had a hundred more Dans in the mobile community, I think we could see mobile technology make the world a better place in our lifetime.

- Scott Weiss: my mentor on all things mobile design. I swear that he has forgotten more about mobile design than I know. Scott's advice helped me get through some of the most challenging chapters to create some of my best work.

- Luca Passani: although you might not always agree with him, Luca is one of the most passionate and spirited people I know. Without him, mobile would just be yet another boring technology.

- Andrea Trassati: there were plenty of times I felt like I was out on a ledge on some of the more advanced topics. Andrea was invaluable in helping me not just to make sense of it all, but to present it to the reader in a way I never could have on my own.

- David Gerton, Jr.: David helped me get through some especially tough spots with additional expertise, research, and writing. Oh, and he helped me with this book too.

- Twitter: it might be odd to credit a website, but my friends, both real and extended, provided me with dozens of resources, ideas, and tips throughout the creation of this book. They helped me write the book they wanted to read. I hope you are happy with it.

- The book staff: Thanks to Steve Weiss for believing in the mobile web and pushing for this book from the start to finish; without his support, this book would never have been written. Thanks to Ginny Bess Munroe, who had to take my ramblings and turn them into an excellent book. Thanks to everyone at O'Reilly who has been so incredible to work with, especially Chris Meredith, who gave me the daily motivation and encouragement to get this book written.

And finally a heartfelt thank you to my wife, Cyndi, who put up with my all of my crazy ideas long before I started in mobile. She, along with my daughter, Penny, manages to unconditionally support and love me no matter what. For that I owe everything.

## CHAPTER 1

A Brief History of Mobile

I like to compare the history of the mobile industry to the work of Umberto Eco: you get what is going on, but it makes your head hurt in the process. The evolution of mobile networks, the devices that run on them, and the services we use every day have evolved at an amazing rate, from the early phones that looked more like World War II field radios to the ultra-sleek fashion statements of today.

If there is one basic principle about mobile, it is that everything is the way it is for a reason. It might not be a good reason, but a reason exists nonetheless. It is the history, or the context, of the medium that gives mobile designers and developers the patience and passion needed to deal with the frequent issues they face in the mobile ecosystem. The mobile industry is a difficult one to jump into without patience and passion.

This chapter briefly discusses the evolution of the mobile medium from the perspective of the device, the most universally identifiable facet of the mobile ecosystem.

In the Beginning

For those of us who are older—that is, over the age of 30—when we think of what a telephone is and try to picture it, we might think of the phone illustrated in Figure 1-1.

The telephone is undoubtedly one of the greatest inventions of mankind. It revolutionized communications, enabling us to reach across great distances and share thoughts, ideas, and dreams with our fellow man, making the world a much smaller place in the process. In fact, the telephone is probably one the most defining technologies of the twentieth century and the most commonly used electronic device in the world today.

For the vast majority of people around the world, the perception of what a "phone" is and what it can do hasn't changed from this iconic image—something you hold up to your ear and talk into—but when those under the age of 20 picture a telephone, they might think of an image similar to the one shown in Figure 1-2.

Figure 1-1. The traditional telephone

Figure 1-2. A modern mobile phone

Although the modern mobile phone is a distant cousin to the telephone, it is a communication *and* information device. It is nearly always connected to the Internet, even if you don't have a web browser open. You can send and receive voice and text messages. You can purchase goods and services without opening your wallet. Plus, it can locate which street corner you are standing on and tell you what is nearby all in a fraction of an instant. Oh, did I forget to mention that you can talk to people, too?

In fact, the modern mobile phone is capable of doing nearly everything you can do with a desktop computer, but with the potential for more meaningful relevance to our daily activities. The mobile phone is not merely a telephone. In fact, modern mobile devices deliver on the long-overdue promises that technology will make our lives easier, but without the cable clutter that drives someone like my wife nuts.

Thinking of mobile devices more as personal computers and less as telephones is a difficult shift in perception. The mobile industry of today has somewhat of a split personality—each side with its own conflicting interests: the first half being the telecom infrastructure and the people who run it, required for everything to work but only focused on the network; and the other consisting of the devices we carry, focused on how and when we interact with the network. And yet a third personality is the Web, the repository of the world's knowledge that we seek to use in the context of our daily activities.

Even the Web is divided within mobile, consisting of the "regular" or desktop web and the mobile web. The desktop web is made up of the sites and web applications designed for a browser running on desktop or laptop computers. In other words, the desktop context involves information that we access typically while stationary and sitting at our desk. The mobile web contains the sites and web applications designed for mobile devices, or the mobile context, which we can access anywhere at any time.

Technically speaking, it is all one Web, at least in terms of the technology that we use to publish information and knowledge. But these two mediums are very different and offer different value to the end user, based on their context—something we will talk about in more detail in Chapter 4.

The Evolution of Devices

Every story has a beginning, and mobile development is no different. Understanding context is something discussed often in this book, and I can think of no better place to start than to go down memory lane and give you the backstory, or historical context, of how we arrived at the mobile technologies of today.

Mobile technology has gone through many different evolutions to get to where it is today. In the industry, we often refer to these evolutions as "generations" or simply "G," which refers to the maturity and capabilities of the actual cellular networks. The cellular network is only one element of the overall mobile ecosystem, something discussed in Chapter 2. To make sense of it all, you cannot rely on the common convention of network generations, as those milestones are too focused on the network and not the true cultural milestone that is shifting how people use technology.

Rather, you need to segment the history of mobile into five distinct eras of devices. Why the device? Over the history of mobile technology, we've seen an interesting phenomenon. Every now and again a device comes along that changes everything. It might not be the fastest or bleeding-edge technology, but it might just be the right solution at the time. It packages all the current capabilities and standards into something people are willing to add to their lives. It opens people's eyes to the potential of mobile, starting with the early days of simply being able to make a phone call from anywhere to today's pocket personal computers.

The Brick Era

The first era I call the Brick Era (1973–1988). My first taste of mobile telephony was with my father's suitcase phone that he purchased when I was still in school—basically a corded receiver connected to a portable radio the size (and weight) of a car battery. My father, never one to throw away a good piece of electronics, took his suitcase phone on a 1998 driving trip of Alaska—along with his more modern and portable mobile phone. Of the two, the suitcase phone was the only device powerful enough to get a signal in the remote areas of Alaska. Nothing is more emblematic of this era than the Motorola DynaTAC (see Figure 1-3) introduced in 1983.

Figure 1-3. The Motorola DynaTAC 8000X was the first mobile phone to receive FCC acceptance, in 1983; DynaTAC was actually an abbreviation of Dynamic Adaptive Total Area Coverage

You might recall seeing one of these behemoths back in the 1980s. They were larger than corded phones of the day and more expensive than using a payphone, so it's hard to believe that Motorola discontinued the DynaTAC as late as 1994. Mobile telephony certainly existed before this device, but for the first time, telephones were cordless and portable.

As noted already, everything in mobile technology has a reason. Brick Era phones required enormous batteries to get the power needed to reach the nearest cellular network site, which back in the 1980s were few and far between.

Brick Era phones proved useful only to those who truly needed constant communication, such as stockbrokers or those who worked in the field, such as salespeople or real estate agents; because they were so enormous and so expensive, they were far too impractical for the majority of us.

In the early 1990s, bulky mobile technology would eventually be added to cars for increased mobility, first as aftermarket devices installed in trunks or under seats and eventually at the factory in luxury automobiles.

After mobile technology started racing down the motorways of the world, more cellular radio towers were needed to provide constant coverage. As more towers went up, the power demands of the devices went down. In other words, the closer you were to a tower, the smaller the device you needed.

The proliferation of mobile technology in this era opened the gates to mobile devices of today. What started as a bulky luxury item became something everyone could fit into their budgets and their pockets.

The Candy Bar Era

The second era, the Candy Bar Era (1988–1998), represented one of the more significant leaps in mobile technology. "Candy bar" is the actual term used to describe the long, thin, rectangular form factor of the majority of mobile devices used during the Candy Bar Era and even today (see Figure 1-4). At this point, network operators started to see the clear value (and big profits) in their burgeoning cellular networks, and a "perfect storm" ensued. The network shifted to second-generation (2G) technology, starting in Finland in 1991. The density of cellular sites caused by increased usage decreased the power demands of the device, making it small enough to fit in your pocket. And finally, with the economic prosperity of the European, U.S., and Japanese middle classes in the early 1990s coupled with the perceived notion of mobile phones being a luxury item, suddenly everyone around the world wanted a mobile phone.

Increased demand meant more competition for providers and device makers, which further reduced costs to consumers. Those in the industry quickly envisioned a future in which everyone had a mobile phone with them at all times.

As the profits from fixed-line telephone operators started to plummet, mobile operator upstarts saw huge financial gains. During the mid-1990s, a mobile device future blossomed in Northern Europe. At the time, a coworker of mine from Scandinavia said to me:

In Scandinavia all the land lines are owned by the government. If you moved, you had to file a request with the government to switch your service to your new address, which would often take months. At one time, I lived someplace for nine months and I didn't have a phone. I actually moved before I ever got service. Then I had to wait again.

So what most people do is they go down to the mobile shop and buy a phone for a few hundred dollars and use it as a primary phone until they get their land line installed, which in my case, took about a year.

Figure 1-4. A Nokia candy bar phone

This era didn't just usher in portability. For the first time, people started to realize that mobile phones could do more than make voice calls. Candy bar phones—so commonly associated with 2G GSM (Global System for Mobile communications) networks—included SMS (Short Message Service) capabilities.

Initially, the idea behind SMS was for the mobile operator to send subscribers a notification of a new voicemail, or other short notifications. But in the early 1990s, due to oversights by mobile operators, text messages were not charged to consumers. Mobile-savvy Europeans quickly realized that they could send messages to their friends for free when voice calls were still fairly expensive by today's standards. Thus, today's abbreviated text language, which is limited to 140 characters, was born.

The Feature Phone Era

The third era, the Feature Phone Era (1998–2008), wasn't nearly as radical a technological leap as the leap from the Brick Era to the Candy Bar Era, but it was an important evolution nonetheless. Up to this point, mobile phones had done three things: make voice calls, send text messages, and play the Snake game. The Feature Phone Era (see

Figure 1-5) opened the floodgates to a variety of applications and services on the phone, like listening to music and taking photos, and introduced the use of the Internet on a phone.

Figure 1-5. The Motorola RAZR, probably the most iconic device from the Feature Phone Era

During this era, GSM network providers added GPRS (General Packet Radio Service), allowing packet-switched data services. This network evolution is most often referred to as 2.5G, or halfway between 2G and 3G networks. Network providers offering CDMA and other TDMA-based networks followed suit with similar packet-switched data services soon after.

With the introduction of cameras into higher-end feature phones and with increased consumer interest in digital photography, demand for feature phones began to increase. Not soon after, we saw the introduction of the Motorola V3, more commonly known as the RAZR. Although the RAZR was not a technologically advanced phone, its slim form factor and sleek appearance drove demand around the world, selling over 100 million units, to become the second-best-selling mobile phone of all time. *PC World* magazine ranked it #12 in their "50 Greatest Gadgets of the Past 50 Years."

I often joke that the camera phone introduced the mobile web to the world, and that the RAZR promptly broke it (referring to its poor mobile web browser), but it achieved enormous market share.

At last, the Web had reached mobile devices, but due to high prices, poor marketing, and inconsistent rendering, no one was using it. Instead, mobile companies were focusing on creating downloadable ringtones, wallpapers, games, and applications to sell through network operator portals.

When I look back on this period, I think about how little real innovation occurred during this time. Everyone knew what the problems were, everyone knew where we needed to go—but no one seemed interested in getting there. It seemed as though mobile insiders just wanted to see quarterly earnings at the sacrifice of the long-term benefit of the medium.

Sure, there were moments of divine inspiration, but they always seemed to get cut short by the demands of network operators and device makers. You simply couldn't innovate within the space without their express permission. With the inconsistent interpretation of agreed-upon standards, consumers felt like spectators of a Wild West shootout.

Hope would come from unexpected places soon, helping to shape a new vision for the future of mobile, but only after a decade of mobile designers and developers shed a lot of blood, sweat, and tears.

The Smartphone Era

The Smartphone Era occurred at the same time as the third and fifth eras and spans from around 2002 to the present. What is and isn't a smartphone (see Figure 1-6) has never been defined, which explains the overlap in chronology. Although smartphones have all the same capabilities of a feature phone, like making a phone call, sending an SMS, taking a picture, and accessing the mobile web, most smartphones are distinctive in that they use a common operating system, a larger screen size, a QWERTY keyboard or stylus for input, and Wi-Fi or another form of high-speed wireless connectivity.

Although there have been many different flavors of smartphone, I see this era more as a technological bridge. Most notable devices of this era seemed to try to be something that they weren't.

For example, the Nokia 9000 series of "Communicator" smartphones looked like a feature phone on the outside, but you when held one on its side, it could be opened like a clamshell to reveal a large screen and keyboard—obviously in an attempt to position the phone as a micro laptop. This is something Microsoft also attempted with its initial Windows CE platform, which would later become Windows Mobile.

Handspring, on the other hand, combined a Palm OS–based PDA with a phone module to create PDA-style smartphones, which would later become the popular Treo line of smartphones. Research in Motion applied its background in two-way paging to create the first BlackBerry, which would later be used to "push" email to corporate citizens in a pager-like fashion.

Figure 1-6. Early smartphones came from companies like Nokia, Handspring, and Research in Motion (RIM)

As you can see, with the exception of Nokia, many smartphone players came from outside of the wireless industry and often found themselves ill-prepared to deal with the increasing demands of network operators and a highly competitive and fast-paced industry. It took several years and many mobile devices for these manufacturers to find the right mix of features and stability.

But even with all this effort, smartphones failed to pique the public's interest and create demand, capturing—at best estimates—just 10% to 15% of the global mobile phone market share.

Meanwhile, feature phone manufacturers continued to evolve their devices, merging the capabilities of PDAs or limited desktop computing with traditional feature phones. Most notably, Symbian—initially a joint venture of mobile device makers Nokia, Motorola, Ericsson, and Psion, and now fully owned by Nokia—created the Symbian OS, a smartphone operating system containing common libraries, tools, and frameworks.

The Symbian OS is used for a variety of mobile devices, the most recognizable of which are the Nokia S60 (still referred to by the defunct Series 60 label), and popular models like the 6260 and N95, both devices that look more like phones than computers.

I think Nokia defined the devices of the Smartphone Era. They created great telephones, an amazing framework for creating cool applications and services, and a reusable infrastructure to innovate.

And although new phones continue to emerge based on the smartphone model, I feel like they will continue to be usurped by the fifth and final era: the Touch Era.

The Touch Era

Change occurs because there's a gap between what is and what should be.

—Craig McCaw

Mobile devices started as simple portable telephones, but they evolved. Messaging was added to mobile capabilities, but mobile devices were still just person-to-person communication tools. We saw networks improve and data speeds increase, which allowed for more technology and more features each year, crammed into smaller and smaller packages. Mobile devices got smarter by learning from desktop computing, truly becoming personal computers, but people weren't interested.

Until recently, the history of mobile has been borrowing from other mediums, learning and growing along the way, but never quite creating an identity of its own. But that all changed. As Steve Jobs once said, "Every once in a while a revolutionary product comes along that changes everything."*

On the morning of January 9, 2007, Steve Jobs went onstage at the MacWorld conference in San Francisco to usher in the fifth and final era and to change the mobile world. He introduced the world to the iPhone.

Now it is at this point that I must confess that I have been a devout Macintosh user for more than 10 years. I also must confess that there are many within the mobile community that disagree with me on the importance of this event, especially my friends from Finland.

I insist that we will look back on the iPhone as one of the most significant milestones that the mobile industry has ever seen. In fact, I believe that in the future, when we reflect on the history of mobile technology, we will divide it into the days before the iPhone and the days after.

Though the majority of technology within the iPhone and its lower-cost successor, the iPhone 3G/3GS, had already been available from several manufacturers for some time, what was so notable about the iPhone was how it changed everyday perceptions of what mobile technology can do. It wasn't a phone, it wasn't a computer: it was something else entirely.

In less than four months after its release in the United States, sales of the iPhone 3G surpassed the long-established Motorola RAZR to make it the bestselling mobile device

---

* *http://www.apple.com/quicktime/qtv/mwsf07/*

in the United States. It also surpassed RIM in quarterly smartphone sales, putting it on its way toward capturing 30% of the overall smartphone market.

Impressive numbers, for sure, but what leads me to believe that the influence of the iPhone and devices like it will have a lasting effect on the future of mobile are stories like this one from the *New York Times* back in March 2008, several months prior to the release of the iPhone 3G:

> M:Metrics, a measurement firm that studies mobile media, has released a survey of iPhone users six months after the device was released to long lines and nearly unending fanfare.
>
> The results, from a January survey of more than 10,000 adults, are somewhat dramatic. 84.8 percent of iPhone users report accessing news and information from the hand-held device. That compares to 13.1 percent of the overall mobile phone market and 58.2 percent of total smartphone owners—which include those poor saps with BlackBerries and devices that run Windows.
>
> The study found that 58.6 percent of iPhone users visited a search engine on their phones, compared with 37 percent of smartphone users in general and a scant 6.1 percent of mobile phone users.
>
> The market for mobile video once seemed like a nonstarter in the United States. Well, 30.9 percent of iPhone users have tuned into mobile TV or a video clip from their phone, more than double the percentage that have watched on a smartphone.
>
> Finally, 74.1 percent of iPhone users listen to music on their iTunes-equipped devices. Only 27.9 percent of smartphone users listen to music on their phones and 6.7 percent of the overall mobile-phone-toting public listens to music on their mobile devices.[†]

Since then, the iPhone's presence on the Web has been increasing twofold per quarter. It now ranks in the top 10 platforms accessing the Internet, above desktop computers that still run Windows 98, Windows ME, and beginning to rival even modern platforms like Linux. Even the iPod touch, which is recognized separately from the iPhone, out-ranks the more-established Windows Mobile and Nokia Series 60 devices, and should reach the top 10 soon.

But the impressive numbers hardly stop there. In less than a year, more than 2,000 mobile web applications were made freely available specifically for the iPhone. Throughout my entire career in the mobile industry, I hadn't even seen 2,000 mobile websites, let alone web applications.

And within just six months of the launch of the iPhone 3G and the ability to purchase and load applications onto the iPhone, the iTunes App Store had already seen its more than 10,000 applications downloaded over 300 million times, at a rate of 2 million per day.

---

† *http://bits.blogs.nytimes.com/2008/03/18/iphone-users-are-mobile-web-junkies/*

"It's unbelievable," says Piper Jaffray's Gene Munster. "It's a differentiator. We think [next year] it's going to be a $1 billion market place and Apple will probably take about 30 percent of that. There's virtually no operating expense for them. They just approve the apps."‡

These staggering numbers aside, the iPhone is just the beginning. Devices that can separate themselves from the clunky smartphones of old, and begin to understand that mobile devices are not just telephones nor miniature computers, but a new medium entirely, will be in a position of strength. The iPhone is to mobile phones as the Macintosh was to personal computers: a market definer. With the iPhone, Apple has set the bar for what people want. The masses have finally realized that a phone is more than just a device that can make phone calls, and they now have expectations about what a phone can be. And they want more.

Mobile devices of the Touch Era are a completely new medium capable of offering real people new and exciting ways to interact and understand information. The devices of tomorrow will be able to leverage location, movement, and the collective knowledge of mankind, to provide people's lives with greater meaning.

And what is so exciting is that "tomorrow" is actually happening right now.

‡ *http://www.cnbc.com/id/28070203/*

## CHAPTER 2

The Mobile Ecosystem

The Internet has spoiled us. We tend to oversimplify the technology powering the Internet. The Internet is actually a complex ecosystem made up of many parts that must all work together seamlessly. When you enter a URL into a web browser, you don't think about everything that has to happen to see a web page. When you send an email, you don't care about all the servers, switches, and software that separate you from your recipient. Everything you do on the Internet happens in fractions of a second. And you have the perception that all of this happens for free.

If you talk to people unfamiliar with mobile, you might find that they quickly assume that the mobile ecosystem is exactly like the Internet, and that all the same rules apply. This couldn't be further from the truth. Mobile is an entirely unique ecosystem and, like the Internet, it is made up of many different parts that must all work seamlessly together. However, with mobile technology, the parts are different, and because you can use mobile devices to access the Internet, that means that not only do you need to understand the facets of the Internet, but you also need to understand the mobile ecosystem.

To put it another way, think of the Internet as a great cloud in the sky. When we want to pull something from it, we use a tool, like a piece of software or device, to interact with it. This can include mobile devices, which we tend to think of as tools. Although this image is partially correct, it's still missing a big piece of the puzzle. To continue the analogy, if the Internet is a cloud, then the mobile ecosystem would be the atmosphere, made up of many clouds, keeping the clouds from drifting off into space; the Internet is just one of these clouds, albeit a very large one.

In case that isn't confusing enough, people in mobile tend to refer to *everything* related to mobile as "mobile." This chapter looks at some of the clouds in the sky and how each part plays into the ecosystem as a whole. It also looks at how you can get started with mobile.

Think of the mobile ecosystem instead as a system of layers, as shown in Figure 2-1. Each layer is reliant on the others to create a seamless, end-to-end experience. Although not every piece of the puzzle is included in every mobile product and service, for the

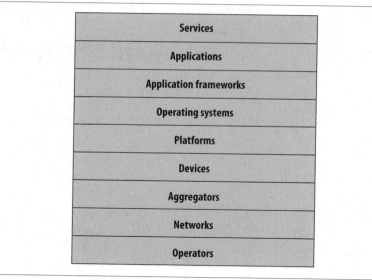

Figure 2-1. The layers of the mobile ecosystem

majority of the time, they seem to add complexity to our work, regardless of whether we expressly put them there.

The following sections expand on each of these layers and the roles they play in the mobile ecosystem.

Operators

The base layer in the mobile ecosystem is the *operator*. Operators go by many names, depending on what part of the world you happen to be in or who you are talking to. Operators can be referred to as Mobile Network Operators (MNOs); mobile service providers, wireless carriers, or simply carriers; mobile phone operators; or cellular companies. In the mobile community, we officially refer to them as operators, though in the United States, there is a tendency to call them *carriers*.

Operators are what essentially make the entire mobile ecosystem work. They are the gatekeepers to the kingdom. They install cellular towers, operate the cellular network, make services (such as the Internet) available for mobile subscribers, and they often maintain relationships with the subscribers, handling billing and support, and offering subsidized device sales and a network of retail stores.

The operator's role in the ecosystem is to create and maintain a specific set of wireless services over a reliable cellular network. That's it. However, to grow the mobile market over the past decade, the operator has been required to take a greater role in the mobile ecosystem, doing much more than just managing the network. For example, they have had to establish trust with subscribers to handle the billing relationship and to offer devices, content, and services that often compete with their partners, who are people like us and who want to create content and services for mobile devices.

Unless you work for an operator, you likely curse their names, at least behind their backs. The operator is viewed as an unfortunate necessity in the mobile world. Often the mobile startups and companies that succeed are the ones with the best "carrier relations man," or person with the best relationship to the operators.

Table 2-1 lists the rank, markets, technologies used, and subscriber numbers for the world's largest operators.

Table 2-1. World's largest mobile operators

Rank	Operator	Markets	Technology	Subscribers (in millions)
1.	China Mobile	China (including Hong Kong) and Pakistan	GSM, GPRS, EDGE, TD-SCDMA	436.12
2.	Vodafone	United Kingdom, Germany, Italy, France, Spain, Romania, Greece, Portugal, Netherlands, Czech Republic, Hungary, Ireland, Albania, Malta, Northern Cyprus, Faroe Islands, India, United States, South Africa, Australia, New Zealand, Turkey, Egypt, Ghana, Fiji, Lesotho, and Mozambique	GSM, GPRS, EDGE, UMTS, HSDPA	260.5
3.	Telefónica	Spain, Argentina, Brazil, Chile, Colombia, Ecuador, El Salvador, Guatemala, Mexico, Nicaragua, Panama, Peru, Uruguay, Venezuela, Ireland, Germany, United Kingdom, Czech Republic, Morocco, and Slovakia	CDMA, CDMA2000 1x, EV-DO, GSM, GPRS, EDGE, UMTS, HSDPA	188.9
4.	América Móvil	United States, Argentina, Chile, Colombia, Paraguay, Uruguay, Mexico, Puerto Rico, Ecuador, Jamaica, Peru, Brazil, Dominican Republic, Guatemala, Honduras, Nicaragua, Ecuador, and El Salvador	CDMA, CDMA2000 1x, EV-DO, GSM, GPRS, EDGE, UMTS, HSDPA	172.5
5.	Telenor	Norway, Sweden, Denmark, Hungary, Montenegro, Serbia, Russia, Ukraine, Thailand, Bangladesh, Pakistan, and Malaysia	GSM, GPRS, EDGE, UMTS, HSDPA	143.0
6.	China Unicom	China	GSM, GPRS	127.6
7.	T-Mobile	Germany, United States, United Kingdom, Poland, Czech Republic, Netherlands, Hungary, Austria, Croatia, Slovakia, Macedonia, Montenegro, Puerto Rico, and U.S. Virgin Islands	GSM, GPRS, EDGE, UMTS, HSDPA	126.6
8.	TeliaSonera	Norway, Sweden, Denmark, Finland, Estonia, Latvia, Lithuania, Spain, and Central Asia	GSM, GPRS, EDGE, UMTS, HSDPA	115.0

Rank	Operator	Markets	Technology	Subscribers (in millions)
9.	Orange	France, United Kingdom, Switzerland, Poland, Spain, Romania, Moldova, Slovakia, Belgium, Liechtenstein, Israel, Egypt, Ivory Coast, Jordan, Cameroon, Botswana, Madagascar, Mali, Senegal, Mauritius, Réunion, Martinique, French Guiana, Saint Kitts and Nevis, Dominica, and Dominican Republic	GSM, GPRS, EDGE, UMTS, HSDPA	111.8
10.	MTS	Russia, Ukraine, Belarus, Uzbekistan, Turkmenistan, and Armenia	GSM, GPRS, EDGE, UMTS	91.7
11.	MTN Group	Afghanistan, Benin, Botswana, Cameroon, Republic of Congo, Côte d'Ivoire, Cyprus, Ghana, Guinea Bissau, Republic of Guinea, Iran, Liberia, Nigeria, Rwanda, South Africa, Sudan, Swaziland, Syria, Uganda, Yemen, and Zambia	GSM, GPRS, EDGE, UMTS, HSDPA, HSUPA	80.7
12.	AT&T	United States, Puerto Rico, and U.S. Virgin Islands	GSM, GPRS, EDGE, UMTS, HSDPA	74.9
13.	Bharti Airtel	India, Seychelles, Jersey, Guernsey, and Sri Lanka	GSM, GPRS, EDGE	72.0
14.	Verizon Wireless	United States	CDMA2000 1x, EV-DO	70.8
15.	SingTel	Singapore, Australia, India, Indonesia, Thailand, Philippines, Bangladesh, and Pakistan	GSM, UMTS, HSDPA	70.7
16.	Telecom	Italy, Brazil, San Marino, and Vatican City	GSM, GPRS, EDGE, UMTS, HSDPA	70.6
17.	Etisalat	Afghanistan, Benin, Burkina Faso, Central African Republic, the Ivory Coast, Egypt, Gabon, Indonesia, Niger, Nigeria, Pakistan, Saudi Arabia, Sudan, Tanzania, Togo, and United Arab Emirates	GSM, GPRS, EDGE, UMTS, HSDPA	63.0
18.	Orascom	Algeria, Bangladesh, Egypt, Pakistan, Tunisia, and Zimbabwe	GSM, GPRS, EDGE	62.9
19.	VimpelCom	Russia, Kazakhstan, Ukraine, Uzbekistan, Tajikistan, Georgia, Armenia, Vietnam, and Cambodia	GSM, GPRS, UMTS	57.8
20.	NTT docomo	Japan and Bangladesh	GSM, GPRS, PDC FOMA, HSDPA	53.5

Although most operators are interested in innovation in the wireless marketplace, they have been known to strangle startups with impossible requirements, such as supporting too many devices or seemingly ridiculous certification processes and bad pricing models.

The days of impossible requirements are changing, however. Today's mobile startups have learned the lessons of the companies that came before them that tried to dance with the devil and lost everything. Many look to the successes of Web 2.0–era startups that were able to start with little infrastructure and quickly grow successful businesses. These startups figured out how to duplicate the phenomenon of mobile, bypassing the operators completely (something this book tells you how to do).

You can compare operators to Big Oil. They both have this thing they know everyone wants, and therefore they can make a lot of money from it. They know they have a limited amount of time to make it. With oil, it is the depleted resources and competition of green energy sources; in wireless, it's the growth of competing wireless technologies, such as Wi-Fi, WiMAX, ultra-wide broadband, and whitespace frequencies.

As competing technologies become more mature, they can't charge as much as they did when they first came out. As consumer options in the market mature, both the oil industry and operators must realize that they can't continue to monopolize their markets. They must realize that they don't control their industries; they are only a player in them. Unfortunately, in the meantime, both of these industries will continue to force us to pay an artificially inflated cost to play.

Networks

Operators operate wireless networks. Remember that cellular technology is just a radio that receives a signal from an antenna. The type of radio and antenna determines the capability of the network and the services you can enable on it.

You'll notice that the vast majority of networks around the world use the GSM standard (see Table 2-2 for an explanation of these acronyms), using GPRS or GPRS EDGE for 2G data and UMTS or HSDPA for 3G. We also have CDMA (Code Division Multiple Access) and its 2.5G hybrid CDMA2000, which offers greater coverage than its more widely adopted rival. So in places like the United States or China, where people are more spread out, CDMA is a great technology. It uses fewer towers, giving subscribers fewer options as they roam networks.

Table 2-2. GSM mobile network evolutions

2G	Second generation of mobile phone standards and technology	Theoretical max data speed
GSM	Global System for Mobile communications	12.2 KB/sec
GPRS	General Packet Radio Service	Max 60 KB/sec
EDGE	Enhanced Data rates for GSM Evolution	59.2 KB/sec
HSCSD	High-Speed Circuit-Switched Data	57.6 KB/sec

3G	Third generation of mobile phone standards and technology	Theoretical max data speed
W-CDMA	Wideband Code Division Multiple Access	14.4 MB/sec
UMTS	Universal Mobile Telecommunications System	3.6 MB/sec
UMTS-TDD	UMTS +Time Division Duplexing	16 MB/sec
TD-CDMA	Time Divided Code Division Multiple Access	16 MB/sec
HSPA	High-Speed Packet Access	14.4 MB/sec
HSDPA	High-Speed Downlink Packet Access	14.4 MB/sec
HSUPA	High-Speed Uplink Packet Access	5.76 MB/sec

Like all things in mobile, we like to merge a lot of technology into overly simplistic terms, which tends to create a lot of confusion. So when we say 3G, for example, we often aren't talking about just the capabilities of the network, but the devices that run on it.

Although the core technology that empowers voice communication has stayed relatively the same, network generations are most often used to describe the data speeds the network is capable of delivering.

Devices

What you call phones, the mobile industry calls handsets or terminals. These are terms that I think are becoming outdated with the emergence of wireless devices that rely on operator networks, but do not make phone calls. The number of these "other" devices is a small piece of the overall pie right now, but it's growing rapidly.

Let's focus on the biggest slice of the device pie—mobile phones. As of 2008, there are about 3.6 billion mobile phones currently in use around the world; just more than half the planet's population has a mobile phone (see Figure 2-2).

Most of these devices are feature phones, making up the majority of the marketplace. Smartphones make up a small sliver of worldwide market share and maintain a healthy percentage in the United States and the European Union; smartphone market share is growing with the introduction of the iPhone and devices based on the Android platform. As next-generation devices become a reality, the distinction between feature phones and smartphones will go away. In the next few years, feature phones will largely be located in emerging and developing markets. Figure 2-3 shows a breakdown of devices.

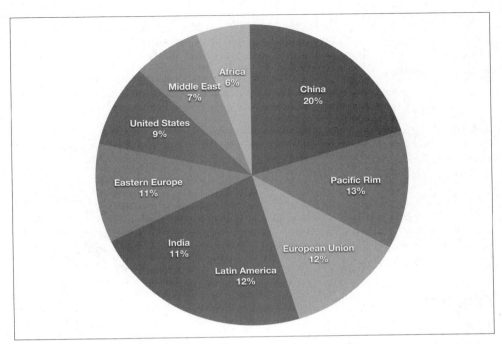

Figure 2-2. Mobile devices around the world

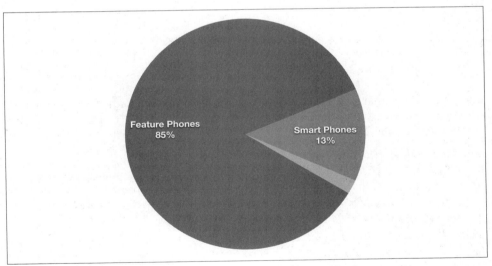

Figure 2-3. Breakdown of devices

Most mobile devices are subsidized in some form or another. Operators sell devices at a severely discounted price, often one-third or less of the actual cost of the device. This enables the operators to lock the devices to their networks. They can then preload onto the device content and services that are beneficial to themselves in exchange for lower price points, encouraging subscribers to upgrade to new devices with new capabilities.

Subsidization means that devices need to be *provisioned* (or customized) to operators' individual requirements. Provisioning dramatically increases the number of devices released every year, with each device being slightly different from the other.

The sheer number of devices is both a blessing and a curse to the mobile industry. On the one hand, the magnitude of the mobile market is huge. It is one of the largest digital mediums mankind has ever seen. On the other hand, so many devices means adapting to those devices—not to mention painful and costly development cycles.

This brings us to the greatest challenge the mobile ecosystem currently faces: device fragmentation, a term used to describe how mobile devices interpret industry specifications differently, causing different mobile devices to display content inconsistently. Despite what you may know or have heard, you can take a deep breath and relax. Device fragmentation is a topic we will clear up completely in the following chapters.

Platforms

A mobile platform's primary duty is to provide access to the devices. To run software and services on each of these devices, you need a *platform*, or a core programming language in which all of your software is written. Like all software platforms, these are split into three categories: licensed, proprietary, and open source.

Licensed

Licensed platforms are sold to device makers for nonexclusive distribution on devices. The goal is to create a common platform of development Application Programming Interfaces (APIs) that work similarly across multiple devices with the least possible effort required to adapt for device differences, although this is hardly reality. Following are the licensed platforms:

Java Micro Edition (Java ME)
: Formerly known as J2ME, Java ME is by far the most predominant software platform of any kind in the mobile ecosystem. It is a licensed subset of the Java platform and provides a collection of Java APIs for the development of software for resource-constrained devices such as phones.

Binary Runtime Environment for Wireless (BREW)
: BREW is a licensed platform created by Qualcomm for mobile devices, mostly for the U.S. market. It is an interface-independent platform that runs a variety of application frameworks, such as C/C++, Java, and Flash Lite.

Windows Mobile

> Windows Mobile is a licensable and compact version of the Windows operating system, combined with a suite of basic applications for mobile devices that is based on the Microsoft Win32 API.

LiMo

> LiMo is a Linux-based mobile platform created by the LiMo Foundation. Although Linux is open source, LiMo is a licensed mobile platform used for mobile devices. LiMo includes SDKs for creating Java, native, or mobile web applications using the WebKit browser framework.

Proprietary

Proprietary platforms are designed and developed by device makers for use on their devices. They are not available for use by competing device makers. These include:

Palm

> Palm uses three different proprietary platforms. Their first and most recognizable is the Palm OS platform based on the C/C++ programming language; this was initially developed for their Palm Pilot line, but is now used in low-end smartphones such as the Centro line. As Palm moved into higher-end smartphones, they started using the Windows Mobile-based platform for devices like the Treo line. The most recent platform is called webOS, is based on the WebKit browser framework, and is used in the Prē line.

BlackBerry

> Research in Motion maintains their own proprietary Java-based platform, used exclusively by their BlackBerry devices.

iPhone

> Apple uses a proprietary version of Mac OS X as a platform for their iPhone and iPod touch line of devices, which is based on Unix.

Open Source

Open source platforms are mobile platforms that are freely available for users to download, alter, and edit. Open source mobile platforms are newer and slightly controversial, but they are increasingly gaining traction with device makers and developers. Android is one of these platforms. It is developed by the Open Handset Alliance, which is spearheaded by Google. The Alliance seeks to develop an open source mobile platform based on the Java programming language.

Operating Systems

It used to be that if a mobile device ran an operating system, it was most likely considered a smartphone. But as technology gets smaller, a broader set of devices supports operating systems.

Operating systems often have core services or toolkits that enable applications to talk to each other and share data or services. Mobile devices without operating systems typically run "walled" applications that do not talk to anything else.

Although not all phones have operating systems, the following are some of the most common:

Symbian
> Symbian OS is a open source operating system designed for mobile devices, with associated libraries, user interface frameworks, and reference implementations of common tools.

Windows Mobile
> Windows Mobile is the mobile operating system that runs on top of the Windows Mobile platform.

Palm OS
> Palm OS is the operating system used in Palm's lower-end Centro line of mobile phones.

Linux
> The open source Linux is being increasingly used as an operating system to power smartphones, including Motorola's RAZR2.

Mac OS X
> A specialized version of Mac OS X is the operating system used in Apple's iPhone and iPod touch.

Android
> Android runs its own open source operating system, which can be customized by operators and device manufacturers.

You might notice that many of these operating systems share the same names as the platforms on which they run. Mobile operating systems are often bundled with the platform they are designed to run on.

Application Frameworks

Often, the first layer the developer can access is the application framework or API released by one of the companies mentioned already. The first layer that you have any control over is the choice of application framework.

Application frameworks often run on top of operating systems, sharing core services such as communications, messaging, graphics, location, security, authentication, and many others.

Java

Applications written in the Java ME framework can often be deployed across the majority of Java-based devices, but given the diversity of device screen size and processor power, cross-device deployment can be a challenge.

Most Java applications are purchased and distributed through the operator, but they can also be downloaded and installed via cable or over the air.

S60

The S60 platform, formerly known as Series 60, is the application platform for devices that run the Symbian OS. S60 is often associated with Nokia devices—Nokia owns the platform—but it also runs on several non-Nokia devices. S60 is an open source framework.

S60 applications can be created in Java, the Symbian C++ framework, or even Flash Lite.

BREW

Applications written in the BREW application framework can be deployed across the majority of BREW-based devices, with slightly less cross-device adaption than other frameworks.

However BREW applications must go through a costly and timely certification process and can be distributed only through an operator.

Flash Lite

Adobe Flash Lite is an application framework that uses the Flash Lite and ActionScript frameworks to create vector-based applications. Flash Lite applications can be run within the Flash Lite Player, which is available in a handful of devices around the world.

Flash Lite is a promising and powerful platform, but there has been some difficulty getting it on devices. A distribution service for applications written in Flash Lite is long overdue.

Windows Mobile

Applications written using the Win32 API can be deployed across the majority of Windows Mobile-based devices. Like Java, Windows Mobile applications can be downloaded and installed over the air or loaded via a cable-connected computer.

Cocoa Touch

Cocoa Touch is the API used to create native applications for the iPhone and iPod touch. Cocoa Touch applications must be submitted and certified by Apple before being included in the App Store. Once in the App Store, applications can be purchased, downloaded, and installed over the air or via a cable-connected computer.

Android SDK

The Android SDK allows developers to create native applications for any device that runs the Android platform. By using the Android SDK, developers can write applications in C/C++ or use a Java virtual machine included in the OS that allows the creation of applications with Java, which is more common in the mobile ecosystem.

Web Runtimes (WRTs)

Nokia, Opera, and Yahoo! provide various Web Runtimes, or WRTs. These are meant to be miniframeworks, based on web standards, to create mobile widgets. Both Opera's and Nokia's WRTs meet the W3C-recommended specifications for mobile widgets.

Although WRTs are very interesting and provide access to some device functions using mobile web principles, I've found them to be more complex than just creating a simple mobile web app, as they force the developer to code within an SDK rather than just code a simple web app. And based on the number of mobile web apps written for the iPhone versus the number written for other, more full-featured WRTs, I don't think I'm alone in thinking this. Nonetheless, it is a move in the right direction.

WebKit

With Palm's introduction of webOS, a mobile platform based on WebKit, and given its predominance as a mobile browser included in mobile platforms like the iPhone, Android, and S60, and that the vast majority of mobile web apps are written specifically for WebKit, I believe we can now refer to WebKit as a mobile framework in its own right.

WebKit is a browser technology, so applications can be created simply by using web technologies such as HTML, CSS, and JavaScript. WebKit also supports a number of recommended standards not yet implemented in many desktop browsers.

Applications can be run and tested in any WebKit browser, desktop, or mobile device.

The Web

The Web is the only application framework that works across virtually all devices and all platforms. Although innovation and usage of the Web as an application framework in mobile has been lacking for many years, increased demand to offer products and services outside of operator control, together with a desire to support more devices in shorter development cycles, has made the Web one of the most rapidly growing mobile application platforms to date.

Applications

Application frameworks are used to create applications, such as a game, a web browser, a camera, or media player. Although the frameworks are well standardized, the devices are not. The largest challenge of deploying applications is knowing the specific device attributes and capabilities. For example, if you are creating an application using the Java ME application framework, you need to know what version of Java ME the device supports, the screen dimensions, the processor power, the graphics capabilities, the number of buttons it has, and how the buttons are oriented. Multiply that by just a few additional handsets and you have hundreds of variables to consider when building an application. Multiply it by the most popular handsets in a single market and you can easily have a thousand variables, quickly dooming your application's design or development.

Although mobile applications can typically provide an excellent user experience, it almost always comes at a fantastic development cost, making it nearly impossible to create a scalable product that could potentially create a positive return on investment.

A common alternative these days is creating applications for only one platform, such as the iPhone or Android. By minimizing the number of platforms the developer has to support and utilizing modern application frameworks, the time and cost of creation go down significantly. This strategy may be perfectly acceptable to many, but what about the rest of the market? Surely people without a more costly smartphone should be able to benefit from mobile applications, too.

Many see the web browser as the solution to this problem and the savior from the insanity of deploying multidevice applications. The mobile web browser is an application that renders content that is device-, platform-, and operating-system-independent. The web browser knows its limitations, enabling content to scale gracefully across multiple screen sizes. However, like all applications, mobile web browsers suffer from many of the same device fragmentation problems.

You could consider the Motorola RAZR to be the epitome of the mobile ecosystem of yesterday. It's been provisioned to numerous operators around the world. It's the perfect example not just of how crazy deploying mobile applications to devices can be, but also of just how bad mobile web browser fragmentation can be. It is a highly prolific device and one that is often recommended for people to support, due to its market penetration. But that is much easier said than done.

If you look at the WURFL database (an open source device repository that is discussed later in this book), you can see that the V3, the real name of the RAZR, has an Openwave 6.2.3.2 web browser. The V3/I/R had the Openwave 6.2.3.4.C.1.109 browser; the V3M/V9M had the Teleca Obigo 4.0 browser; the V3X had the Openwave 6.2.3.1.C. 1.112 browser; the V3M had the Openwave 6.2.3.1.C.1.115 browser; the V3XXI had the Opera 8.0 browser; and the V8 had the Opera 8.5 browser. This isn't even half the list!

From the consumer and business perspective, these are all Motorola RAZRs. But in terms of supporting the RAZR, these might as well be seven different devices. Each of these RAZRs carries very different versions of common applications, each customized for the operator on which they are intended to work.

When a device is sold to an operator, it is provisioned (customized) to their requirements. This means the operators will often put customized applications on each of the devices sold. With the example of the RAZR, every operator had it and every operator put a different web browser on it. To make matters worse, the RAZRs, like most phones, are not field-refreshable, meaning that you can't update the software, upgrade the applications, or eliminate bugs.

For example, if a device manufacturer makes a device called the MDv1, they must strike a deal with an operator if they want to preload an operator store application, a different web browser, and bowling game. The device is sold as the MDv1.1. The operator sells the devices, or worse, gives them away for free. A couple hundred thousand of them go out into the marketplace before a glitch in the hardware is detected, such as dropped calls. Because the device cannot be upgraded by cable or over the air, the operator stops selling the MDv1.1, but seeing that they have a hit, they quickly replace it with the MDv1.1.1. The whole process is repeated as it is provisioned to each operator. Suddenly, there is an MDv1.2, an MDv1.3, an MDv1.4, and so on. Then we have the next generations—the MDv1.2.1, the MDv1.3.1, the MDv1.4.1, and so on, spreading like a virus. This is essentially what causes device fragmentation, making application development a costly and timely endeavor.

Services

Finally, we come to the last layer in the mobile ecosystem: services. Services include tasks such as accessing the Internet, sending a text message, or being able to get a location—basically, anything the user is trying to do.

What makes the mobile environment such a complicated space to design and develop for are these layers, which the user must wade through in order to accomplish a simple task like "I want to send a text message," "I want to get on the Web," and "I want to access Google." The user has so many opportunities for failure that creating a valuable experience is virtually impossible.

How do everyday people use their mobile devices? What are their impressions of the mobile web? Here is some of the consistent feedback you might hear if you ask these questions:

"It's crap."

"I don't use the mobile web, just because it's awful."

"It costs too much."

"I don't know where my browser is."

"I don't know how to enter the URL."

"I want to go to Wikipedia, but I don't know what to do."

"How do I check my email?"

That is, of course, only if you are old—past your early 30s. The younger you are, the more likely you are rely on mobile services for daily information. Earlier generations—those born since the birth of the Internet—have a unique talent for being able to figure out complicated informational spaces. They are more patient with technology and more apt to explore new methods of accomplishing tasks.

And although one day the youth of today will inherit the digital world, for the time being, the mobile ecosystem is a complicated, fragmented, political nightmare. If I were an entrepreneur looking to create mobile services, knowing what I know about mobile, I would run away, fast. I would probably open a restaurant, which would likely have a higher chance of success.

But we've already seen the future of mobile development, in the form of the iPhone. The iPhone attempts to solve many of the problems facing the mobile ecosystem, from how people interact with their phones, to where we buy our phones, to what type of applications we will pay money for, to the level of technology standards we can support on constrained devices. What makes the iPhone special is how it attempts change on virtually all fronts, something no other device, or company for that matter, has been able to do previously.

Now Apple has done it, the gates are wide open for anyone. People in the industry aren't as jaded anymore, and there is a feeling of excitement and optimism. Although many of the problems in the mobile ecosystem are yet to be fixed and we still have plenty of nonsense to contend with, we can see the light. We can see the path to innovation, to creating applications and services that can quite literally reach the entire planet and quite possibly change the world.

It begins here, with you, right now. Today is the first day of the future of mobile.

## CHAPTER 3

Why Mobile?

It is hard to think about mobile development apart from the buzz it generates. I wish I could tell you that the majority of mobile strategies start with a well-thought-out plan of how to use the medium to meet the needs of users or further the goals of the business. The people at the majority of companies I visit, those I meet at conferences, those writing articles online, and even a few mobile experts claim that mobile is the next big thing, but few can explain why. This is something I have struggled with myself over the years.

We know that mobile devices are proliferating around the world like mad. We know we can perform a lot of different tasks with them. But what makes them the killer device of the next year, decade, or generation? If you've been in the business as long as I have, you know that mobile has been the "next big thing" for at least the last 10 years. We've gone through crazy ups and downs and seen periods of massive innovation, followed by depression, then innovation, then depression; the sad cycle repeats itself.

To date, the majority of the mobile industry outside of the operator has been unable to sustain a true long-term growing business. Instead, it has jumped from one bubble to the next, trying to time the leap so as not to be the one left to pop it. This is the perfect example of an investment-funded industry. Being unable to create a long-term sustainable business model on its own, innovation relies on selling the "next big thing" to investors instead of selling what people really want.

To understand how to design and develop mobile products that benefit users, and therefore long-term business, you need to answer the question, "Why Mobile?" Unfortunately, you can't just answer the question with one simple and pithy answer. Mobile is not only a new medium, but also a new business model entirely. There is opportunity aplenty, but the trick is learning how to harness the market to the benefit of the business.

Therefore, I suggest that you look at mobile in the same terms that visionaries like Richard Branson and Steve Jobs might look at it: building a successful long-term business around the underserved needs of real people. Invest in the opportunity to capitalize on the current needs and define what the market can be.

Being a visionary doesn't have to be rocket science. Taking the leap into the mobile market doesn't have to be risky; it simply requires an honest look at what exists, what users want, and then taking the next logical step. This chapter helps you answer these questions:

- What is the size and scope of the overall mobile market?
- What is the addressable market, or what percentage of the overall market will we be able to reach?
- What are the competitive benefits that the mobile medium has to offer?
- What core needs does our addressable market have? Or, for the uninspired, what horrible products are serving the market that people are "putting up with"?
- Based on past trends and market needs, where is the market headed in the near term? And is that in line with the long-term market direction?
- How can we cost-effectively address the needs of the market and the needs of the user? (The answer might surprise you.)

Size and Scope of the Mobile Market

The earth's population is a little over 6 billion. To break it down, in Figure 3-1 you can see the sizes of various countries around the world.

The United States of America has a population of 303 million people. The European Union's population is 495 million. India's is 1.2 billion. China's is 1.3 billion, roughly one-fifth the population of the planet. If you compare these numbers to the number of mobile devices in Figure 3-2 you see some startling numbers.

Over 3.6 billion people own or have access to mobile devices today. Of those, over 1.6 billion (or 25 percent of the world's population) have access to the Web through a mobile device—a number that is growing with each year. What is interesting and un-expected is that just 1.1 billion people have access to Internet-connected desktop computers.

Mobile devices have already outpaced the majority of media we rely on every day, including computers. Today, more people access the Web via a mobile device than via a computer, and the disparity between the two numbers will grow more severe in the coming years, as shown in Figure 3-3.

The sheer size of the mobile web—the largest, most available mass medium to mankind—is good justification to create mobile products, or at least to define a mobile strategy. Most companies start and end at this point.

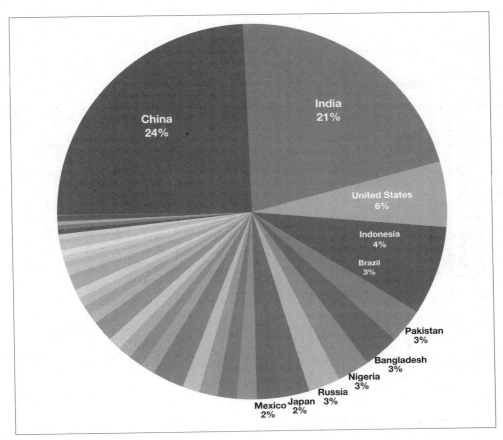

Figure 3-1. Global population

The Addressable Mobile Market

There should be no doubt that the overall mobile market is enormous. The question is, "How many of these users will our product be able to reach?" Unfortunately, this is difficult to answer, because no one knows for sure.

The market isn't unlike the early days of the Web, a kind of lawless wild frontier in which no one cared about the addressable market; given that the cost of publishing was close to zero, anyone could experiment with the medium. In the mobile market, the cost of publishing can go from zero to $60,000 in a blink of the eye, so understanding the likelihood of a return on the investment is important.

Many companies fall into two broad categories: either they have tried something in mobile and were not impressed with the results, or they want to try to do something in mobile but are wary of making the investment needed.

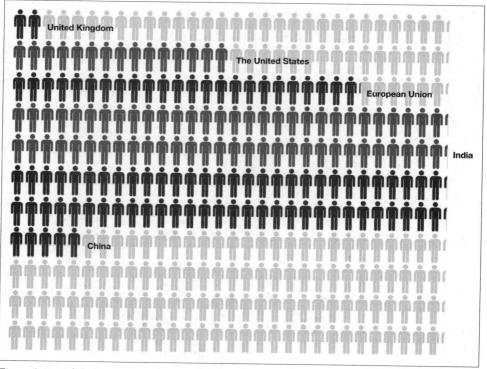

Figure 3-2. Mobile population

There are research and analytics firms that do the research and charge a hefty price tag for their findings. Although this data is useful at times, to foster innovation and increase exposure to addressable market segments, and ultimately drive usage across the board growing the market, you need to make reasonable assumptions about the market with freely available data.

Because the costs can increase so quickly as more devices are supported, you must look for ways to keep publishing costs as low as possible by targeting the most suitable mobile device. To do this, break the market down to four comparisons: high-end versus low-end devices, best-selling versus free devices, mobile web versus native applications, and touch versus D-pad (directional pad) devices. These are discussed in the following sections.

High-End Versus Low-End Devices

What is considered to be the high-end device at any one time? What makes it the best in its class? Is it the technical features or design features that appeal to users? Can you

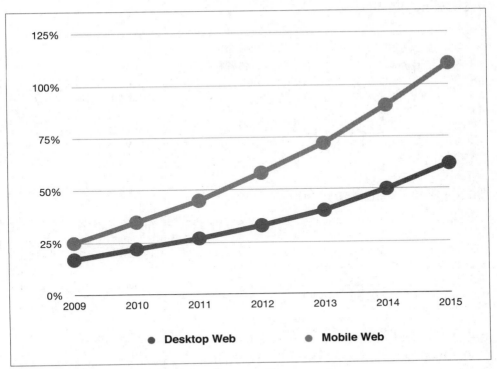

Figure 3-3. Predicted growth of mobile web access

compare this device to the low-end devices? Is it the low cost or the simplicity of features that appeals to users?

Go to your local mobile phone store and pretend to shop for a new phone (resellers provide you with a lot more information than an official operator's retail store). I find the geek on the street a far better source of research than a pricey market report. If you get to know the salespeople a bit, they will spill all sorts of tasty information your way—plus, you won't have to pretend to be a shopper all the time.

Ask them about what features people are asking about—the camera, the media play-back features, or the web browser? Find out what drives shoppers into the store, and find out what features motivate them to come in and how these compare to those on the devices they actually end up with.

You can also play the "I'm buying a phone for my parent/spouse/daughter/son" scenario, describing your target market as a family member. The salesperson will tell you exactly what your target market is buying and why. The main goal is to determine the breakdown of high-end to low-end devices that are sold to your target market. This percentage will become useful later.

Best-selling Versus Free

This comparison is similar to the high-end versus low-end comparison: ask what shoppers are buying this month and why. What is the most popular device in the store when window shopping versus the most popular device people leave with? Comparing and contrasting these two different ends of the device spectrum tells a lot about the addressable market. It can tell you what people want versus what people are willing to spend. You can find out whether they are skeptical about mobile technology or just incredibly cost-conscious. Again, try to find out what your market is going toward and why. If you can get a breakdown of the most popular devices, you are walking away with gold.

Mobile Web Versus Native Applications

Look for data online. Seek out any press releases or free data reports that discuss mobile web usage versus native application downloads. These numbers are incredibly hard to come by, so look for fragments of information and try to assemble a trend. Supplemented with a little geek-on-the-ground device research, you can come up with the percentages on each side, as the type of device purchased by the target market goes a long way toward determining what they will be able to do on their device.

Touch Versus D-Pad

Finally, find any data, online or from your new retail friends, on what the usage breakdown is between top touch devices and the more traditional D-pad devices. There is certainly more of a trend toward touch devices that have far more features than the traditional devices. Knowing what mode your target market wants to use to add and receive content will tell you a lot about your users and the content they will want to consume or interact with.

When you put the results of this research together, you can gain a unique view of your target market, what features they want, and what they have. You will know what your users want to do with mobile technology versus what they are willing to pay for, if you will need to sell your product through an operator or app store, or if you can distribute it yourself. Plus an astute listener will hear what people want versus what they put up with, a critical piece of information for rising visionaries.

Mobile As a Medium

After examining the size and scope of the mobile market, you need to understand what it means to your users and ultimately how it will benefit the business.

One of the best introductions I've ever heard was from Tomi Ahonen, an author and expert on next-generation wireless, who describes mobile as "the seventh mass media." He points out that each of the mass media we use every day has advantages and disadvantages, each playing a significant role in society, with mobile being the latest in the series.

The following sections briefly summarize each of the seven mass media. By better understanding what the users are accustomed to and what purpose media plays in people's lives, you can better understand how to leverage the mobile medium to suit your goals.

The Printing Press

The first mass medium was the printing press, one of the greatest inventions of mankind. The time needed to publish information was dramatically reduced, enabling information to be copied and distributed farther and faster than handwritten predecessors. Not to mention there was less damage from time or the elements.

The printing press has continually played a crucial role in history. For example, during the American Revolutionary War, the printing press was used to mass-produce the record of civil unrest occurring in Boston, the epicenter of the colonial uprising. It is hard to imagine that the Continental Congress would have had the public support needed to form the United States without the aid of the printing press and people like Boston printer Samuel Adams.

Recordings

The second mass medium was the recorded sound, initially on Edison's phonograph cylinder and later on more durable materials like glass, vinyl, magnetic tape, or today's compact disc. Although we normally associate recordings with music, early recordings enabled people to share information by hearing it firsthand and recreating the experience by sharing it over time and over great distances.

But recorded music also played an important part in influencing society. Jazz in particular gave new opportunities to freed slaves in America as entertainers. Only a generation after the end of slavery, African-American jazz musicians became popular figures in modern music—thanks in large part to their recordings.

Cinema

The third mass medium was the cinema. Like recordings, we tend to think of cinema as entertainment, but cinema enabled a visual experience to be shared over time and distance. Suddenly, we were able to witness distant or past events firsthand, enabling the viewer to draw conclusions from what he or she saw and heard.

My father is old enough to remember watching the newsreels of World War II. Seeing dramatic events from halfway around the world in his local cinema had a lasting effect on him and his entire generation, rallying nations toward the conflict.

Radio

The fourth mass medium was radio—an extension of recordings, but including the live broadcast of material. Information could be distributed as it happened and as far as the radio signal would reach. Like cinema, radio could give listeners an intensely personal experience. And because recording technology was becoming smaller, events could be recorded where film cameras could not go.

Think of Winston Churchill's radio addresses, which helped bring hope and confidence to the people of Great Britain during the frequent air raids of World War II. Or, think of Edward R. Murrow's radio reports from the battlefield, which brought the war into living rooms around the world.

Television

Television is the fifth mass medium. The early days of television were more of a visual extension of radio. As the price of televisions dropped and they entered more homes, television transformed itself into a more iconic medium—one of the most influential and certainly one of the most disruptive. The television became a more practical alternative to previous media like cinema and radio. Suddenly, we could participate with information in more intimate and visceral ways.

When I think of the influence that television has had on culture, I think of how events like landing on the moon, the conflicts in Southeast Asia, the Beatles performing on the *Ed Sullivan Show*, Elvis Presley's dance moves, and even the weekly adventures of the *USS Enterprise* transformed culture. How we perceived the world and each other changed drastically in a fairly short amount of time because of the television.

The Internet

Nothing seemed to happen for a long time after the invention of the television—that is, not until we started plugging our computers into the phone jack to hear that weird tone that meant we were connecting to the Internet and the World Wide Web. And we all pretty much know how that went.

Among the important early Internet developments were the dot-coms, which gave us a reason to have a computer in our home and not just at work. Then came Web 2.0, which showed us that the Web could be used in meaningful ways. It took a little time to find its path, but eventually the Internet would become a transformative and disruptive mass medium in its own right. The Web of today is threatening the printing press and crumbling newspaper empires that have been around for a century.

The iTunes Music Store, which sells digital content over the Internet, is now the largest purveyor of recordings in the world. You can purchase, download, or stream movies through many Internet-connected devices. Podcasts and streaming audio are transforming what "radio" means. And today, almost all major television networks are either selling or streaming their content online.

But what the Internet cannot do—and what four out of the previous five media can do—is be portable. Of course, we have laptops and Wi-Fi, but those technologies are local, offering only a small number of clients with a broadcast distance of a few meters. This brings us to the seventh mass medium.

Mobile

The seventh mass medium, of course, is mobile technology. The mobile industry actually started around the same time as the Web, but it took it years for us to consider it a mass medium. The mobile medium is actually quite deceiving; it would be easy to see it as an extension of the previous media, but mobile is actually quite unique, as the only mass medium that can do everything the previous six media can do.

Understanding how the mobile medium stands apart from other media is an important step in determining how to best leverage mobile for business goals. The following sections discuss not just how real people can use mobile technology in place of previous media, but also the unique and competitive benefits.

Read and publish

Reading text is consistently one of the most frequently performed tasks on a mobile device, second only to making and receiving phone calls. From sending an SMS (Short Message Service) or email message to reading some news or even a book, we are increasingly using mobile devices beyond just communication with one another and now for distributing knowledge and for absorbing information.

New technologies like Amazon's Kindle provide a new method of reading, both on a specific reading device and on devices like the iPhone. By using mobile technology, we can synchronize our reading position among contexts; browse, purchase, and download content over the air; and even interact with our friends and our reading groups, giving our thoughts and opinions, regardless of location.

Play recordings

The iPod is still the most popular portable media device around the world, but that hasn't stopped mobile device makers from including media playback features in just about every new mobile device. We're certainly seeing that people don't want to have to carry multiple devices around if they don't have to, which explains part of the rapid popularity of the iPhone.

The opportunity for sharing recordings over the air with friends, or by proximity, provides a new kind of word-of-mouth advertising, one that could revolutionize the recording industry.

Watch movies

For the cinephiles, watching a movie on the small screen of a mobile device may sound sacrilegious, but I can't tell you how many times having a few movies on my smartphone has come in handy. From boring cross-country flights to long road trips with my school-age daughter, having a little entertainment on hand can be a refreshing filler for idle periods.

But it goes beyond just watching movies: imagine watching trailers for movies showing at nearby theaters, or finding out at which nearby theaters your friends are in line getting tickets. Or even go beyond traditional entertainment: imagine watching clips of events happening around you, or maybe interacting with your social network in real time as you watch videos wherever you are. The opportunities are endless.

Listen to radio

Many mobile devices include a radio tuner to play broadcasts from the local radio waves in the air. But we can also stream live broadcasts from the BBC, National Public Radio, and others, getting information as it happens from the source we want, regardless of our locale. In addition, we are seeing a rise in the popularity of personal radio services like Pandora, which allows you to create unique radio stations based on an artist or song and then stream it to your mobile device.

Imagine a day where you can listen to a service like Pandora in your car, where content is streamed over the air or locally from the phone in your pocket directly to the in-dash entertainment system. Imagine a day where you can get real-time personalized updates, like a kind of audio RSS, where you can stay up-to-date while stuck in traffic.

Watch television

Many operators use the ability to watch live television on a mobile device to highlight the capabilities of their mobile broadband networks, though it has been slow to catch on with consumers. In countries like Japan and South Korea, seeing someone watch television on a mobile device is a commonplace occurrence; in other countries, not so much. However, it is still early. The television networks are only now starting to understand and extend their business to the Internet. Mobile technology won't be far behind.

Like radio, the opportunity for using the always-on connection and the personal nature of the handset means that a television broadcast like news could become contextual, relevant, personalized, location-based, or maybe based around the interests of an individual's network.

Use the Internet

Of course, without the Internet, many of these services wouldn't be available on a mobile device. And as discussed previously, access to the Internet and usage of the Web from a mobile device are increasing exponentially every month around the world.

Mobile's Unique Benefits

If it wasn't enough that mobile can do everything the other mass media can do, as Tomi Ahonen points out, mobile has five unique benefits that none of the others do:[*]

"The first truly personal mass media"
> For example, we don't usually share our mobile devices with our spouses. Each of the other mass media are or can be shared easily; even your computer can be shared at home or at work. Mobile devices offer us, for the first time, a means to interact with information in a personal and intimate way.

"The first always-on mass media"
> Many don't realize that mobile devices have the capability to send and receive information at all times, even when idle. Unless it is turned off, your mobile device is connected to the network. Although it may sound Big Brother-ish, this can be a huge benefit to the average user, enabling the device to predict tasks based on your location and anticipate the information you will need based on your surroundings.

"The first always-carried mass media"
> How many other mass media can you think of that we carry with us at all times? Seven out of ten people sleep with their phone within arm's reach. A colleague of mine doing some user research on how people use mobile devices even found that the place in our homes that we most frequently use mobile devices is the bathroom. Maybe we need to separate ourselves from our phones a bit more!

"The only mass media with a built-in payment channel"
> Every phone sold from an operator has a built-in means of purchasing content, even goods and services known by the funny acronym of BoBo, or Billing on Behalf of. You pay for it and it is charged to your phone bill. When you factor in the statistic that twice as many people have phones than credit cards, suddenly the potential seems enormous.

"At the point of creative impulse"
> We are able to create content and distribute it the moment the mood strikes us. From taking a picture of something interesting and uploading it to social networks in order to share it with our friends to capturing a video of an important event and sharing an experience, mobile devices enable us to create and publish in near real time. Information and experiences can now be shared with audiences around the

---

[*] Sources: *http://communities-dominate.blogs.com/brands/2007/02/mobile_the_7th_.html* and Tomi Ahonen, *Mobile as 7th of the Mass Media* (futuretext).

world as they happen and from multiple points of view. It is simply unlike any other medium we have ever seen before.

The Eighth Mass Medium: What's Next?

The argument of the seventh mass medium makes a strong case for how the mobile medium is unique and powerful, but any good investor might ask what is next. I suggest that we are at the beginning of the eighth mass medium, which I'll call "ubiquity."

Ubiquity includes ubiquitous computing, ubiquitous network, and ubiquitous media. We are moving toward an age where the Web, together with mobile technology, can create a write-once-publish-everywhere environment. When we merge the portability of mobile technology, pervasive always-on networks, and the proliferation of web standards and services across media, we begin to see a picture where everything starts to converge.

As discussed previously, operator control and device fragmentation are some of the most significant and costly hurdles to mobile design and development. I've worked with many mobile companies over the years, and each of them has been forced either to bend to the will of the operator or to deal with porting applications and services to hundreds of devices. These two factors alone put many innovative companies out of business, because they simply couldn't afford to get their products to market.

Even today, creating a mobile application for multiple devices delivered through multiple operators in just a single market can easily cost two to three times more than it would cost to create a more full-featured desktop application.

The concept of a ubiquitous medium has the potential to change that situation, starting with the mobile web. I say "potential" because we aren't quite there yet, but we are close. The mobile web of today is based on the same standards as the desktop web. In the very near future, as mobile and desktop browsers get smarter, the thin line that separates them will disappear entirely.

It is possible today to sit down and create a mobile website for millions to use, regardless of the device or operator. All the user needs is a mobile web browser, which is available in virtually all modern handsets, and an active connection to a cell tower. This ease and pervasiveness is an impossibility for mobile applications that depend on the device API and must be approved and distributed through the operator, or in the case of smartphone platforms like the iPhone or Android, through an external application store.

There are still many challenges of compatibility across multiple devices in today's mobile web, namely JavaScript and Cascading Style Sheet (CSS) support, which produce inconsistencies when users view the same code across multiple browsers. But luckily the mobile web is based on standards, specifications, principles, and best practices agreed upon by the mobile community. These standards are nearly identical to the standards that govern the desktop web. The primary difference is that not all mobile browsers have implemented them yet. But they are close.

To put this into context, I can create a working mobile web application based on desktop and mobile web standards that works on iPhone, Android, and BlackBerry platforms; plus, it will support all the devices that run Opera Mobile, like the Nokia S60 platform. By the end of this book, you will be able to do it, too. If we were creating native applications, it would take half a dozen developers and months to do it.

You can take ubiquity one step further, beyond mobile devices. Services like Twitter, which blend the Web, mobile and desktop applications, and widgets around a single service. Using the same standards, techniques, and principles, we can create experiences for browsers, mobile devices, native applications, widgets, e-book readers, in-dash systems, gaming consoles, media centers, and so on.

The scope of this book can't possibly cover all of these outlets, nor should it. It is still early. But it all begins with the mobile web, the first phase in a long list of platforms for which our content can be formatted.

Ubiquity Starts with the Mobile Web

We have endured years of bold and usually unfulfilled claims that come from the tech sector. We've been promised that the Web will make our lives easier, but aren't we seeing the opposite reaction? Our lives are becoming so infused with information that it becomes overwhelming and even stressful just to keep up—an increasing problem called *information overload*.

The problem: the Web of today is wide, but not deep. Although we have access to enormous amounts of information, the majority of it isn't meaningful. It lacks depth and value for our lives. For example, according to an October 2008 Nielson report, the average person in the United States looks at 76 web pages per day, spending an average of 55 seconds per page. The short duration suggests short information-gathering tasks, idle browsing, or a more severe problem with the Web: distrust.

Web content expert Gerry McGovern describes it this way:

> People are more skeptical about content online than offline. People basically view the Internet as a dumping ground for content. There's some great stuff, sure. However, it is vastly outweighed by badly written, out-of-date, inaccurate, and sometimes deliberately misleading content.

With the mobile web, however, we see a quite different picture. We still see short, simple tasks that one might expect given the mobility of the user, but with a far higher number of page views per visit, each with longer session times. People don't just bounce around from site to site, they invest time and absorb information.

In the mobile web, the top content segments continue to be news, email, weather, sports, city guides, and even social networks. People seek out information relevant to their task or location, adding valuable personal context not often found on the desktop web.

The mobile web has an amazing capability to add context to information, adding immediate relevance to what we are doing right here, right now. This unique benefit makes the mobile web stand out from all other media as a means of finding and participating with information in intensely personal ways. Figure 3-4 illustrates this point.

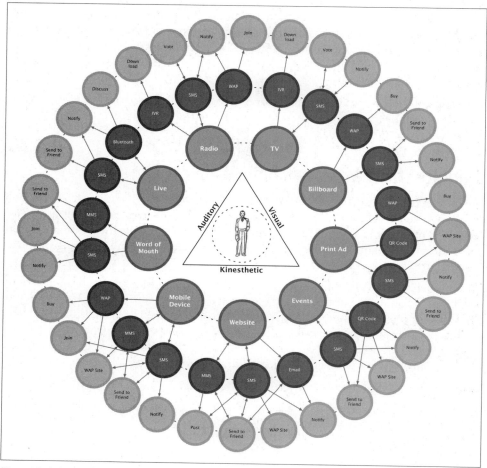

Figure 3-4. Seeing how users interact with the mobile web

Imagine the person in the middle of all the information. He is surrounded by the three modes through which we process information: visual, auditory, and kinesthetic (touch). Directly around us we have information media like the Web, print, and so on. Each of these can leverage different mobile technologies (the middle ring) available in every mobile device to empower the user to interact with the medium in very personal ways (the outer ring).

With just a few short actions, people can perform personal and contextual tasks using a device they most likely have on them at that exact moment. This not only changes our behavior to be more in tune, aware, and informed of our surroundings, but it also creates entirely new markets.

Tomi Ahonen describes the impact mobile devices have on humanity this way:

> The [mobile] phone is bigger in its reach than the car (800 million), TV (1.5 billion), or Internet (1.1 billion). It will make bigger changes in the next decade than any of these did. The phone adds the combined utility of the fixed telephone, Internet, computer, credit card, and TV. The phone will impact your life in more ways than we can imagine, because of its multi-functionality aspect, and its reach.

It all starts with the mobile web, the only ubiquitous platform available on all mobile devices. We are at the precipice of the next generation of the Web, not the mobile web, and not the Web as we know it today, but an entirely new web. These two forces of nature are converging in a novel and interesting way to create something new. Nobody can claim to know where it's going to go, but we definitely know that it's happening.

We're seeing the mobile infrastructure grow at a phenomenal rate, picking up speed in innovation and deployment. We are seeing the proliferation of mobile devices growing, especially higher-end devices such as the iPhone and the Android platform, which bridge the divide between mobility and computing.

A political quote, though technically unrelated, comes to mind from the 2008 U.S. presidential campaign. After the Republican party was delivered a decisive defeat in the executive, legislative, and state races, Jill Lawrence wrote in *USA TODAY*, "If the Republicans don't make their peace with Hispanic voters, they're not going to win presidential elections anymore. The math just isn't there."[†] Lawrence was referring to the rising Latino population in the United States, expected to triple to 30 percent of the U.S. population by 2050.

What does this have to do with the mobile web? To rephrase the sentiment: people who think they can ignore the mobile web and succeed by launching new brands and services on the desktop alone need to wake up. The math just isn't there.

---

† *http://www.usatoday.com/news/politics/election2008/2008-11-06-hispanics_N.htm*

Think of the future of mobile technology this way. On May 6, 1998, the first iMac was announced. It was one of the most significant influences on the consumer perception of desktop computers in its day. The first-generation iMac had a 233 mHz processor, and 4 GB of storage. Nine years later, on January 9, 2007, the iPhone was introduced. It was one of the most significant advancements in the consumer perception of mobile devices and the mobile web. The first-generation iPhone had a 412 mHz processor and 4 GB of storage. In less than 10 years, the technology that powered the first iMac became the technology that powered the first iPhone. Now imagine what the next 10 years will bring!

## CHAPTER 4

Designing for Context

In late 2008, I was in Berlin doing a mobile workshop at the Web 2.0 Expo. Having never been to Berlin, I did what I always do in new cities that I visit—I explored. I enjoy just walking aimlessly around a new city with no particular destination or plan. Not only is it a relaxing way to see the sights, but I find amazing things that aren't on any tour or in any guidebook. There was just one problem with my plan: everything I explored was in German.

Because I know only about five and a half words in German, this made exploration more challenging. Although I really enjoyed seeing Berlin, my first of couple days there were an empty experience. I could certainly see what was right in front of me, but I didn't have any understanding of what I was looking at. I found myself constantly wondering, "What is this place? Who is this a statue of and why is it here? What significance does it play? What are these people doing? Why is this important?"

At the end of my first day, I found myself in my hotel room visiting Wikipedia so that I could read about what I had seen that day. Unfortunately, I couldn't remember all the sites I had seen. The next day was my workshop, so I had only a little time to wander, but I started taking photos of plaques and historical markers with the camera in my phone so I could translate them later that night. I thought it was an inspired idea, but it proved to be too difficult to make out the text and then enter it into an online translator. My third and final full day in Berlin I had the entire day to explore. To make the most of it, I completely gave up on my futile attempt to avoid incurring international data roaming charges and started using my iPhone as my own personal tour guide.

I wandered aimlessly like I did the previous days, but this time, as soon as I found myself in front of an interesting cultural landmark, I pulled out my phone and started using the location features of the device to detect my location and show me information, in English, about the nearby sites and how far away they were from my precise location. I would read the history of the landmark, what it meant to the German people, and learn about other nearby landmarks that were similar. My day of sightseeing, the locations I visited, and the experiences I had were defined by the information I received on my mobile device.

There I was, the geekiest tourist you've ever seen, standing in front of a great historical landmark, looking down at my phone reading the Wikipedia entry about it (see Figure 4-1).

Figure 4-1. My mobile device enhanced my understanding of the landmarks in Berlin, in my own language

It might have appeared stupid to others and maybe it made me stand out as a tourist; if it did, I didn't care. My experience of Berlin was immensely richer because I had an understanding, or context, and I was experiencing it in my immediate surroundings. The irony is that as a "mobile guy," it would take me several months to realize that my experience was almost the perfect example of mobile context.

Thinking in Context

Context is probably the most used, underestimated, and misunderstood concept in mobile. I think of it as the chewy nougat in the center of a good candy bar. Sure, the candy bar would be good without it, but it's that little extra bit that makes the candy bar an incredible experience; no one quite knows what it is, but everyone knows that it tastes good. Actually, that example probably doesn't help demystify context as much as it makes you crave a candy bar. So let me explain it another way.

I define context in two ways. There is "Context" with a big C and "context" with a little c. These are often used interchangeably with no preference or distinction.

Although they are the same word, they have two different implied meanings. This isn't to say that the case of a letter makes one more important than the other. It only helps to make a distinction between lofty big ideas (big C) and the more practical and more invisible intention of use (little c).

Context with a Capital C

Context with a capital C is how the users will derive value from something they are currently doing, or in other words, the understanding of circumstance. It is the mental model they will establish to form understanding. For example, standing in front of the remnants of the Berlin Wall and reading about the history on my phone is adding Context to my task. It enhances my experience and awareness of my surroundings in a significant way.

I refer to this as "providing Context" as in "this information is providing me Context or better understanding of what this moment in time means to me." Context enables us to better understand a person, a place, a thing, a situation, or even an idea by adding information to it.

The distracted driver

One example of putting Context to use is the Distracted Driver campaign that ran in New Zealand.* The government saw an increasing problem with drivers having "diverted attention," causing as many as 26 fatalities in a single year. Research found that tasks such as receiving an incoming text message seriously diverted drivers' attention and increased the chances of an accident. In an effort to build awareness, the New Zealand Transport Authority asked Clemenger BBDO to develop a campaign to demonstrate that sending text messages while driving can be dangerous.

Do you remember hearing the stories of those gory highway safety scare films from the 1960s, such *Options to Live*, *Wheels of Tragedy*, or *Mechanized Death*? It used to be that you had to show teenagers extremely violent and horrific images to illustrate the importance of safe driving. But Clemenger BBDO was able to use Context to prove a point instead of blood, guts, and dismemberment.

The result was a website showing a driver casually behind the wheel, almost as though you were looking at a live webcam from inside a car somewhere in New Zealand (Figure 4-2).

* http://www.youtube.com/watch?v=EUgq_h8clv8

Figure 4-2. The distracted driver

You are then encouraged to send a text message from your phone to the driver seen in the website (Figure 4-3).

Figure 4-3. People are encouraged to send a text message to the distracted driver

In a few seconds, the driver in the video would receive the text message, picking up his phone to read the message you just sent him (Figure 4-4).

Figure 4-4. The distracted driver receives the text message

As the driver (and you) is reading the message, he slowly heads into oncoming traffic and crashes (Figure 4-5).

This merger of mobile, web, and video technology creates a dramatic and compelling case for when you should read your text messages. By using Context to form understanding—in this case, of the situation of being behind the wheel, followed by an event (the crash) that would be impossible to duplicate in real life, the user is able to participate in a hypothetical situation in real time. The user becomes part of the experience. Without Context—in this case, the ability to interact with the driver using our mobile phone—our understanding of what transpired would be diminished. We would simply be watching one of those shocking highway safety films, with no personal or emotional stake in what just happened.

Figure 4-5. The distracted driver crashes due to your text message. Nice work!

The eRuv project

Another fascinating example is the mobile eRuv project (*http://www.dziga.com/eruv/*), which Elliot Malkin created in New York City. An *eruv* is a wire boundary that often surrounds orthodox Jewish communities (see Figure 4-6). Every Saturday, the Jews observe the Sabbath, or the day of rest. During the Sabbath, Jews are unable to perform any work of any kind; for those that observe the stricter concepts, this means that you can't carry anything outside of your private domain, such as carrying your house keys outside of your house. The eruv designates a conceptual area within a public space, as a shared private space for the community where Jews can still observe the Sabbath, carrying their keys with them from their home, but without committing sin.

Fifty years ago, a large area of Manhattan was designated as an eruv without wires, using the Third Avenue elevated train on the West and the East River on the East (Figure 4-7).

Figure 4-6. A modern-day eruv in the Upper West Side of Manhattan

Figure 4-7. The pre-1955 eruv, using physical and natural landmarks instead of wires

In 1955, the train line was dismantled and replaced with a subway, and thereby the eruv was dismantled, too. To experiment with these concepts and what a modern eruv might look like, Elliot used QR, or quick response, codes (also known as semacodes in the United States) to designate the boundaries of the old Third Avenue eruv (as shown in Figure 4-8). QR codes are two-dimensional barcodes that can be read by the cameras within mobile phones, and display text, or contain a URL to a mobile web page or photo. In this case, Elliot showed a picture of what the eruv looked like 50 years ago.

Figure 4-8. An eruv built with QR codes, instead of wires, showing 50-year-old location information

The eRuv project shows that physical locations can contain information, too. In this case, location contains the information long since gone with the passage of time; it is shown using mobile technology that can be extracted from a mental model of how space is perceived around us.

Context with a Lowercase c

On the other hand, context with a lowercase c is the mode, medium, or environment in which we perform a task or the circumstances of understanding. The following sections look at the three types of context with a little c.

The present location or physical context

Let's start with physical or environmental context. Where I am, my location more often than not has meaning. My physical context influences my actions; whether I'm at home, in the office, in a car, on a bus or train, walking the streets of Berlin, etc., each environment will dictate how I access information and therefore how I derive value from it.

The easiest example might be using a mobile device in a car versus a bus or train. My car is private, so the information I want to extract might be far more personal in nature, but I can't easily input or extract information from a screen because I'm focused on not running off the road and ending up like the distracted driver. On a bus or train, I have little or no privacy, so the information I seek will likely be casual and less meaningful, but because I have no other tasks to perform, I can focus on using the device.

One interesting mashup of the real world augmented by information from the Web is the Android application Wikitude (*http://www.mobilizy.com/wikitude.php*), which allows you to use the camera in your phone and literally overlay information about the landmarks in front of you, as shown in Figure 4-9.

Figure 4-9. The Android application Wikitude presenting information from the Web in the context the user's present location

Our present device of access, or media context

Next let's talk about media. We talked about mass media earlier and the various benefits each medium offers. What we didn't discuss is why people use one medium over another and how each medium provides users different levels of value. For example, a printed newspaper is rich in content, but offers little value to our current task or location. A newspaper is a fixed artifact of the past, given the time it takes to print and distribute. By the time you read it during your morning coffee, it is already out of date.

Mobile devices may not be as rich in content, but they can provide information on the present, which is a lengthy way of saying that the mobile context offers value that the printed context simply cannot. However, media context isn't just about the immediacy of the information we receive—it also can be used to engage audiences in real time, something that other mediums just can't do.

The Estonian government will be putting the concept of media context to the test in their 2011 parliamentary elections, allowing citizens to vote for their leaders using SMS. In this case, the government can tabulate results instantly. But imagine a day when citizens can vote on local or national issues in real time, eschewing having to wait for traditional media to report on the effect of their vote, instead seeing the results in real time, as it happens.

There are already many opting to use the mobile media context in order to be heard. On the immensely popular television show *American Idol*, more votes were cast using a mobile phone in 2009 (178 million total text message votes) than votes cast in the 2008 presidential election (131 million ballots cast).

Our present state of mind, or modal context

The third and final type of context with a little c is modal context, or our presence of mind when we perform a task. Now this one is by far the hardest to explain. Against my mother's better judgment, I decided I would be a geeky technologist when I grew up, rather than a brainy psychologist, but I'll try to explain it as best I can.

Each time we are exposed to various stimuli, our brains go to work trying to understand the signals our senses send to our brain. Through some miracle of the central nervous system, these signals are converted into reactions. Sometimes they are involuntary physical reactions, like pulling your hand away from a flame, and sometimes they are mental reactions, which we call thoughts and reasoning, like asking yourself, "Should I cross the street now or later?"

We go about our lives and make thousands of little decisions every day that we don't even consciously consider—perhaps habitual or perhaps instinctual. We make maybe a dozen or so inconsequential decisions like "What should I eat for lunch?" Things that seem important at the time, but we promptly forget. And maybe once or twice a day various decisions string together to form a bigger, more meaningful thought such as "Should I buy it now or later?"

As humanity becomes more infused with information and ubiquitous access to it, for better or worse we begin to rely on information to aid us in making these decisions, usually for the more complex ones, and increasingly for the more minute.

Our present state of mind is probably the greatest influencer of what, when, and where we perform tasks. Driven by need, want, or desire we make choices that attempt to accomplish goals—sometimes lofty ones, but more often than not trivial ones. This modal context is at the heart of all deliberate action or inaction.

This concept is in desperate need of an example, so let me use my trip to Berlin once more. While in Berlin, I wanted to gain understanding of how a single city came to be so emblematic of the cultural division that for over half century influenced the rest of the world, and more importantly how the city overcame these differences to become something new.

Is there something I can learn from the transformation of Berlin culture and apply to my own life? Is there something I can teach my child, who will otherwise read about the events that occurred in this great city only in history books?

My presence of mind was to somehow transform artifacts of events and of a people into some type of information that is relevant to me in my own life. Presence of mind drives our actions, to consciously or unconsciously create circumstances in which we gain understanding through the acquisition of information.

Taking the Next Steps

Wow, that was a little painful, but at least I made my mother proud. All right, let's put all the head-shrinking behind us and get to the practical stuff. The reason to dive into all this big C/little c nonsense is to illustrate that what has meaning to the user and what has meaning to us as the creators of experiences is different in terms of context. Can you guess which one is which?

You guessed it: users really don't care about physicality, media, or modes, but that doesn't mean that we can ignore them. Just the opposite—good design means that these elements should be invisible to the user. It is our job to create intuitive experiences using technology to anticipate and solve problems for the user through the fewest, deliberate actions. Users tend not to care about what they can't see, which works out for us because making great experiences pays our bills.

Mobile devices, unlike any other medium, present an amazing opportunity to create contextual, meaningful experiences unlike anything we've ever seen. The trick is mastery, which like all good things, requires time, patience, and understanding. If you can unlock the state of mind of your users and start thinking in their context, understanding how a mobile experience will add value to their lives, you will have the ever-elusive "killer app."

Do this by thinking of your apps in the different contexts your users will likely encounter:

- Who are your users? What do you know about them? What type of behavior can you assume or predict?
- What is happening? What are the circumstances in which they will best absorb the content you intend to present?
- When will they interact? When they are home and have large amounts of time? At work, where they have short periods of focus? During idle periods, while waiting for a train?
- Where are they? Are they in a public space or a private space? Are they inside or outside? Is it day or is it night?
- Why will they use your app? What value will they gain from your content or services in their present situation?

- How are they using their mobile devices? Are they held in the hand or in the pocket? How are they holding it? Open or closed? Portrait or landscape?

If you have ever studied journalism, you might notice that the previous is actually the inverted pyramid of telling a news story, expressing who, when, where, what, and how in the first paragraph. The intent is to define the circumstances, or rhetoric, of how we communicate and understand ideas. Interestingly, these concepts go back thousands of years to the works of philosophers Aristotle (384–322 BC) and later Cicero (106–43 BC).

These are not exactly new ideas; we only need to apply them to the medium at hand (no pun intended). Context is what makes mobile such a powerful and exciting medium. It takes the foundational concepts of western civilization and evolves them to create a new means of communication and understanding, or what ancient Greeks called *inventio*, meaning both invention and discovery.

## CHAPTER 5

Developing a Mobile Strategy

I've been trying to stay positive and paint mobile design and development in a positive light. Before I let you in on the secrets of the mobile field—getting into all the dirty details, pitfalls, and frustrations of the medium, and how to solve them, or in some cases how to sidestep them—I want you to remember how important mobile development is to the future. It is the greatest communication and information medium of our time.

You will be tested. In fact, you should expect your explorations into mobile to be some of the most challenging moments in your career. Never give up: it is worth every minute of frustration, and with a little practice, I think you will find it highly rewarding work that transcends just mobile apps or sites.

It starts with developing a mobile strategy. By the phrase "mobile strategy," what I mean is "how much time, effort, and money it will cost you." Have no illusions: mobile design and development don't come cheap. If you target only one platform, you might possibly see success with relatively no risk. However, once you start attempting to scale that strategy, you will quickly find yourself exponentially increasing the amount of time and effort required to be successful. With the wrong strategy, it can be hard to justify and even harder to monetize the costs of mobile.

Of course, there are workarounds, but if you come from the world of web products or traditional software development, some of the choices you will have to make might be difficult. I can't tell you how many people I've talked to over the years who had a great idea for a mobile product, but didn't see it through because their decisions were guided by the traditional rules of business on the Web or the desktop.

You could almost look at your mobile design and development strategy like a game of *Tetris*—the game where you need to fill in each row with the falling shapes in order to clear the row and continue to make space for more shapes—which happens to be one of the most popular mobile games ever sold. When the game starts, clearing the first few lines is insanely easy. This can be your first hurdle in the mobile field, such as defining your business goals. Your second task gets a bit harder, such as defining user goals, but you are still in control. As the pieces continue to fall, and begin to involve defining the context, design challenges, framework issues, and testing, you start to struggle with getting each of them lined up and placed into the right spots, but you still have a strategy in mind.

Then you start getting into those little S-shaped pieces, or the devices. You aren't sure what to do all with of them, but they keep coming and you aren't able to figure out how to make them all fit into your strategy. Suddenly you realize that you aren't even trying to clear lines anymore—you are just trying to manage the chaos.

I can guess what you are probably thinking: that sounds awful. But let's not forget that *Tetris* is one of the best-selling games ever made, so obviously it can't be that bad. It is certainly frustrating at times, but it is still fun. I think the problem many people have when trying to create a successful mobile strategy is they look at the endeavor like a project, with a predefined beginning and end and a linear path between.

Instead, look at mobile development more like a puzzle, similar to *Tetris*. Sure, it can probably be solved, but investing ages to solve a puzzle does not unlock the great secrets of life. The goal of a puzzle, the value we derive from it, comes from the attempt to solve it—not always from the solution itself. A good mobile strategy is not always just about simply doing something from end to end, as the complexity of supporting the vast number of mobile devices is out of reach for the great majority of us.

I recommend that you view your mobile strategy as a movement that can be transformational to your project, company, or organization. It is about discovering how to infuse a new medium into your business and build an innovative platform that will take you not just into the next year, but into the next decade. For example:

- For a sales-based organization, you could use the mobile web to get information to your salespeople in the field, allowing them to bring up live pricing and estimates while they are at lunch with a client.

- For health care, you could use mobile devices to provide records; access to formularies, policies, and procedures; even patient charts—and all at the patient's bedside.

- For real estate, you can provide potential buyers with live listings of similar properties nearby, information about the house, school information, the average home values in the neighborhood, and mortgage calculators—all while touring a property.

- For local governments, you can use mobile technology for mass transit, to increase awareness or community participation, to elicit feedback on public works projects, and to weigh in on important issues facing the community.

- For retailers, you can provide instant points of purchase without customers having to go to the register, provide ongoing customer support, link online customer reviews and price matching information—all while the customer is holding a product in her hand.

- For arts and entertainment, you can provide information (or even games) about upcoming events and other background information that is all tied into an address book, allowing the customer to plan a fun night out with friends.

The list could go on and on. You might notice that the common thread in each of the opportunities listed here is putting the user in the center of the experience, which is what mobile technology does. Empowering your users empowers the future of your business. And at the center of that transaction is this little always-on, always-connected device that we call the mobile phone.

New Rules

Mobile is a different medium and is governed by a different set of rules. Great mobile products are able to adapt or even look beyond traditional strategy to identify new and unique ways to address both the challenges and the benefits that the mobile medium has to offer.

I've seen many well-funded, well-staffed, and well-intentioned companies fall from great heights because they followed the wrong rules. I think it has given me somewhat jaded insight, as I constantly ask myself, "Why do I think I can succeed where companies with loads more resources failed?" With this in mind, I present to you what I consider to be the new rules for creating a mobile strategy.

Rule #1: Forget What You Think You Know

The first step is to set aside what you think you know. I've talked to several executives and even developers who are quick to explain that they ignore mobile because of this reason or that reason. Their reason is usually based on outdated or even incorrect information that is guiding their assumptions about the medium.

The truth is that the mobile industry is a highly competitive one, with many big companies and many investors and stakeholders. This produces an enormous amount of press releases filled with speculation and empty promises. With so much hot air being released, some is certain to make its way to the tech journals, blogs, or a casual chat with a knowledgeable technologist. Remember the massive amounts of speculation during the dot-com boom? The mobile industry makes the early dot-coms look like amateurs of spin in comparison.

Mobile development is an ever-evolving ecosystem with many facets. Even people like myself who have been in the industry for a while can keep up with only a small fraction of the bevy of new announcements and technologies that are announced in any given week.

Do yourself a favor and forget everything you think you know about mobile technology. Start at the beginning with your project. Ask yourself or your team the hard questions about your business, about your users, and about your development capacity, unfettered by the latest hype, tool, or technology.

Following are some tips and tricks to help you forget what you know:

- Leave your baggage at the door. Forget what you think you know about mobile. It is most likely incorrect.
- Don't try to emulate what has been done before and put that in front of your user. Focus on what is right for your user, not what is right for someone else's user.
- Start at the beginning. Even if a project has been in development for a while, you can still start fresh. Provide new perspectives, and breathe life into the project right from the start.

Rule #2: Believe What You See, Not What You Read

In mobile, any argument can be made, and for a few thousand dollars you can buy a report or white paper that supports your argument. This doesn't make an argument true, but it certainly can be convincing. And with so many people looking at mobile as an additional revenue stream, but completely clueless as to its inner workings, people are grasping for any information they can find—however inaccurate.

Over the years, I've seen the industry sway from opportunity to opportunity to no avail. Each year is named the year of something. Five years ago it was the "year of messaging." The next year it was video, then mobile social networks (cutely labeled MoSoSo, for "mobile social software"), then it was the mobile widgets, then it was the return of the downloadable app (popular eight years ago), then back to messaging. It is maddening.

There are several firms that provide detailed analysis and insight into the mobile ecosystem, but their data is either very costly or too abstract to create buy-in, forcing us to deduce our own conclusions anyway.

I know there are all sorts of scientific methods of conducting market research, but I've found the most foolproof means of research to be simply talking to people about what they use and how they use it, balanced with your instincts regarding what you think the market will buy.

Don't perform the cardinal sin of focus groups, which is asking them to validate your own ideas. Talking to users is about gaining insight into and understanding their daily routines. How and when do they use their mobile devices? What tasks do they perform with them? What information is most valuable to them? What influences their decisions? Understanding your market is understanding your users. Understanding what influences their decisions will tell you what influences their actions.

Don't forget to record everything. Any argument to stakeholders becomes a lot more convincing when you are able to play for them precisely what your users say they want.

Remember that this is a new medium and open for exploration, for innovation. Experts and pundits tend to get excited about a lot of things, but have very short memories.

They pay their bills by making lots of definitive statements, like "you must do x because users expect y." Or "the iPhone taught users to a, so you must do b." With regard to that last statement, many people counted Palm out of the mobile business after the iPhone was released, but then Palm released webOS, which is frankly one of the most innovative uses of mobile and web technology ever seen in mobile. And who knows if it would have been possible without the iPhone. The point is that there is plenty of room to explore and innovate with mobile technology, just as we saw with the desktop web.

Or as Daniel Appelquist, one of the W3C Mobile Web chairs, puts it, "The only prevailing wisdom in mobile is that there is no prevailing wisdom."

Following are some tips and tricks for believing what you see and not what you read:

- Don't trust any report, fact, or figure that is more than a year or two old. It is most likely wrong. For example, the majority of assumptions about mobile development pre-iPhone are no longer applicable.
- Perform contextual inquiries, not focus groups. Go to your users and ask them questions in person, in their context, not yours. They often have a lot to say; listen and keep an open mind.
- Record everything. Nothing makes your case like your users' own words. They have a funny way of reducing company politics and focusing back on the user.
- Don't forget to innovate. Try new things, be bold, and don't be afraid to fail.
- The best strategy succeeds even if it fails. Have a contingency plan. If your plan fails to meet expectations, how can you reuse what you've learned or done on something else?

Rule #3: Constraints Never Come First

This is the tricky one, even for the most seasoned mobile expert. Mobile is a highly constrained medium with many technical obstacles. It is hard not to have constraints of the medium, like devices, networks, or frameworks, influence your decisions.

Creating a great mobile strategy takes a passion for research, some vision, a little risk, and an optimistic, even foolhardy belief that you can make a difference. Unfortunately, these are not the traits you typically find in the halls of corporations. The larger the organization, the more difficult the challenge. We will get to how to circumnavigate this shortcoming later, but for now just try not to cripple yourself from the start.

Thinking big isn't easy. There is a saying in Japan: "The nail that sticks out gets hammered down." There are plenty of people willing to shoot you down for no reason. It only takes one experienced mobile expert or one technologist with something to prove in a brainstorming meeting to completely deflate the vision and strategy definition process. Sorry, guys—you're awesome, but it's true. These people love pointing out technical constraints and, because they are probably the same sort that Rule #1 was written for, you have double the challenge.

Trying to teach the importance of context will get you only so far. In my experience, most great ideas get shot down for the simple reason that someone somewhere didn't feel he was included, so he sabotages movement by calling up petty constraints and presenting them as deal breakers. If you need this constraint-focused group on your side, call a meeting, but avoid having anyone's boss present, at all costs. This might mean having a casual lunch and asking the group for their opinions.

Just start with a big idea, unfettered by constraints, and allow the brainstorming process to run its course. You probably already have a beginning, middle, and end to your project already in your head, but listen to your team and hear out their concerns. You need the support of every single resource you can get on your team. Darker days aren't too far ahead.

I sat in on a brainstorming meeting for the early planning for a mobile product. The project leader called some of the most creative and best thinkers in the company together to come up with some ideas on how to develop a mobile strategy. The leader had done a lot of research on their current traffic and what devices were hitting their site. She had put together some user personas, or fictitious examples of the target customers. Things started slow, with the group initially talking about what had been done by some of the competitors: what had lacked any vision, and what was poorly executed on. After a while the ideas started flowing from the group. Initially the ideas were lofty, but eventually they started to coalesce into something interesting.

Throughout the process, we talked about the users—who they are, what they want, how they are going to use the content we can provide to make their lives easier. We started whiteboarding, just sketching out the ideas. Now although I can't reveal the name of the company to you, I will tell you that the idea that was being formed was one of the better ideas I had heard. It was a great idea. This concept had the opportunity to dramatically improve the bottom line of the organization; define them as a leader in the space; and dare I say, redefine an industry for the next generation, just by using mobile technology to do something useful for people. The energy in the room, as you might imagine, was palpable. Everyone knew we were on to something that could be huge. I tell you this because what happened next completely blew me away.

In came one of the leaders of the company. He was invited to participate in the meeting, but of course got held up in something else. As he walked in, you could practically hear the excitement in the room get sucked out the door as it closed. The leader apologized for being late and asked everyone to resume. We made a feeble attempt to get back on track before the project leader stood up and started to make an ad hoc presentation to the executive, knowing that we had hit the jackpot and proud of what she had accomplished. He listened to the presentation, and others starting chiming in, confirming just how great of an idea it was.

Then for no reason at all, the executive started to kill the idea. He talked about his own experience at a previous company in a different industry, how it didn't work then, so it would not work now. Then he tried to point out several technical concerns, which were quickly put down by the development leads in the room. Then he talked about how this wasn't part of the direction of the company. We tried to counter with our evidence to the contrary. Then he talked about how the project leader really needed to run these ideas up the chain to several other leaders, because this should really fall in their domain, as if we had no right to even discuss the problem.

It was madness. Here was a team fired up, ready and willing to innovate. They had put together a great idea that met and exceeded the customers' needs, that integrated the mobile medium into the broader business strategy, and was frankly one of the best ideas—and it was getting squashed by this executive for arbitrary reasons. If it was anyone else making those remarks, she would have been shot a disapproving glance and a "you aren't helping."

In the end, the project was never started; it died while it was still on the vine. Most of the folks in that room eventually transferred to other departments or left the company altogether. And as far as I know, the executive who killed the project is still there.

The moral of my story: don't invite *that* guy to the room. Mobile projects can be hard to kick off. Start with the big ideas and don't let the many constraints of the medium kill your project before you've begun.

Following are tips and tricks for how to make the constraints not come first:

- Don't invite that guy to an early-stage brainstorming session. Leave him out of the picture until you have something more concrete that will get him excited about the idea. If you don't know how to do this, go to the people that have been around for a while and ask. They will most likely know how to get those guys excited about your idea.

- Hopefully it doesn't need to be said, but don't *be* that guy. Refer back to Rule #1 and forget what you think you know. It's probably wrong, and your team is trying to make you even more successful. Let the team do this.

- If you are concerned about the constraints of the mobile medium, know that there will always be constraints in mobile. Get over it. It isn't a deal breaker. Just make sure you aren't the deal breaker. Focus on strategy first, what they user needs, and lay down the features; then, if the constraints become an issue, fall back to the user goals. There is always an alternative.

Rule #4: Focus on Context, Goals, and Needs

We already discussed context abstractly. By now you understand that context is crucial to creating any mobile product strategy, but how do you find it? When designing a mobile strategy, predicting the user's context is very hard. It is a moment in time, with so many variables. We can mentally picture only so many circumstances that could impact the equation before our heads feel like they're going to explode. It isn't impossible, but it isn't easy.

Although product managers might enjoy trying to plan out every conceivable variable at the end of a complex chain, filling our inbox with specifications, user stories, and edge cases, let's start at the beginning of the chain and reverse-engineer it. Assume for a second that the user's context is based on action. If actions stem from goals and goals are a manifestation of needs, then the user's needs would be the root of the user's context. Needs, unlike context or even goals, are something we can safely predict by looking at some basic information about our user.

Let's use a simple example here: everyone has to eat. It is a basic physical need that everybody must fulfill on a daily basis. The goals associated with this need could be a variety of things, like wanting a particular cuisine, eating only healthy foods, and what to shop for at the local market. The context adds even more variables to our goals, like where I am and what is available around me.

In building a mobile product strategy, we could start with context and create a simple mobile restaurant directory that shows you restaurants based on current location. This would no doubt be a valuable tool for some, but only for a particular set of users that has that particular need at that particular moment—essentially, in that context. In order to get users to actually use your app, you need to sell the importance of that moment and then you can sell the value of your app.

On the other hand, we don't have to sell that people need to eat. We know the need exists in everyone, so what if we built our product based on needs? We'd obviously build our directory in a very different way, starting by structuring our data around various goals. Some users want a particular cuisine, some want to eat healthy, and some want to make a simple meal at home. We can then add the users' context to make that data more meaningful. Here are the nearest Thai restaurants. Here are the calories for a variety of Thai dishes. Here is a recipe and shopping list for a simple Thai dish and the nearest markets with the ingredients.

We never tell users what they can do or limit their options based only on their context. Just the opposite: the users tell us what they want to do and then we filter it based on their context, adding information to their present circumstances.

This isn't to say that drilling down from situations won't produce the same results as building up from needs. But by focusing on needs, we empower the user to perform a task she is likely to do anyway and our product is immediately positioned as something of value her mind.

Following are tips on how to focus on context, goals, and needs:

- Defining the users' context is the first thing to do. Without it, you don't have a mobile strategy; you have only a plan of action.
- Uncover the users' goals, and then try and understand how the users' context alters their goals.

- With goals understood, figure out the tasks the users want to perform.
- Look for ways to filter content by context, such as location, media, and model.

Rule #5: You Can't Support Everything

One of the first steps in building a mobile strategy is taking an honest look at what devices you can support. I'm here to tell you that you can't support everything.

There are literally hundreds of various device models sold around the world each year. And there are dozens of browsers, each with their own quirks. Think about testing a desktop site for Internet Explorer 6, a web browser notorious for its awful web standards support but immense in its market share, meaning that you have to do a lot of things wrong in order to serve the greatest slice of the market. Mobile development has the same challenge—but don't forget to multiply the pesky browsers by a factor of 10. Supporting and testing for even a small percentage of the devices in the market will bury your resources before you've got your mobile product to market.

Start by targeting not just the devices you know your users have, but the ones that you know are consuming lots of data. Your first step will be to decide what you'll support and what you won't. And yes, it is OK to support just one class of device, as long as you know you have the users to back you. The trick is knowing the right device to support.

I know that everyone wants an iPhone site, and maybe you should have one, but is that the device that is going to best serve your customer? If you are targeting stay-at-home moms, then maybe it is a Motorola RAZR. If you are targeting businesspeople, then it is probably a BlackBerry. If you are targeting creative types, then it is probably correct to assume use of an iPhone.

The best place to start is by looking at your server access logs, which will tell you the top browser user agents that are accessing your site currently. These are the users that are seeking out your content, even though it isn't optimized for a mobile device. These devices should be your top priority, enabling you to take advantage of the currently most popular devices for your site. Then look for devices that render content somewhat similarly.

Alternatively, use the niche nature of mobile device marketing to your advantage. Operators often sell devices to specific consumer segments—a social network-enabled phone for the teenager, a camera phone for parents, a simple phone with a big screen for grandparents. If you aren't sure what devices constitute what niche, take a trip to your local operator store and pretend to shop for a phone for your wife, mother, child, coworker, or whomever describes your target audience. The salesperson will tell you exactly what is the most popular device for that market segment.

However you decide to find out your target device, start simply, with the devices that reach your core market. See how your users respond, and only then start to plan the next phase of your project and determine which additional devices you can support.

Here are some tips and tricks to remember when it comes to device support:

- Don't kill yourself by trying to support everything. Start with the devices that best represent your core customer.
- Remember, the most popular device or the one that's easiest to develop for might not always be the best device for your project.
- Check your server logs for the devices currently accessing your site. These are the first devices to target.
- Go to your operator store and do a little market research to find out the recommended devices for your target customer.

Rule #6: Don't Convert, Create

The question I am most frequently asked is "How do I convert my product to mobile?" and my answer is that you don't. Simply porting a site, application, or game to the mobile medium is a big mistake. Although there is plenty that we can learn from products in other media, the mobile market is unique in its challenges and its benefits. The best place to begin a mobile strategy is by creating a product, not simply trying to reimagine one for small screens.

Remember the early days of website design? We had this incredible new publishing medium, and as more people got online, the need for websites and web pages increased. Many looked to print designers to help create web content, often making sites that were more like glossy brochures than the web pages of today, filled with images, animations, and even sound. They took ages to load, and the text was too hard to read and difficult to search through. It took years to undo the numerous bad habits that we so quickly adopted—and quite nearly killed the Web before it even began. It was all because we designed by the rules of the wrong medium.

Great mobile products are created, never ported. Start by understanding your users and the benefits the medium has to offer. Integrate needs, goals, and context and you are off to a better start than most.

Take an airline website, for example. Simply taking the web experience and trying to put it on a small screen doesn't help the user at all; in fact, it has the opposite effect. If the user is on the way to the airport and needs to check whether a flight is delayed, the last thing your user has time to do is scroll around to find where to check flight times. If you've found yourself racing to make a flight and needing to find your flight information, such as times, gate, etc., you need that information quickly.

Creating, rather than converting, experiences specifically for mobile devices allows your users to get information that is faster, friendlier, and more accurate.

Here are some tips and tricks for how to create instead of convert:

- Understand your user and his context. Having an idea of how and when users will access your content will aid in understanding how to best create a tailored mobile experience.
- Don't forget that mobile is a unique medium with its own benefits. Don't try to simply apply the same rationale to your mobile strategy as you would your web or print strategy.

Rule #7: Keep It Simple

And if there was only one rule that I could impart to you, something that would summarize all of the previous rules, it would be this one: keep it simple.

Mobile devices are simple, but not stupid. In other words, they are actually very intelligent computers, but people want to use them in a simple way. People don't want you to offer all the features of your existing application or web app. They want simple features that address basic needs and nothing more. The user must deal with many constraints; therefore, we need to show restraint in the mobile products we build.

Adding feature after feature is an easy trap to fall into, but keeping things as simple as you can, from the structure to your design and the devices you support, has numerable benefits. Staying simple means that you'll have far fewer problems from start to finish, making it easier to get it to market, and meaning you'll learn more from your users sooner, in turn meaning that you can iterate and evolve your product faster, more cheaply, and better than others.

Tips and tricks include:

- Seriously, keep it simple! It can be a big challenge, especially for larger organizations, but try to limit the features to only those that are most crucial to your users. Never put your corporate goals or objectives before the users' interests.
- Try to determine the need that will motivate users to act or interact, and build the experience around that and nothing else.

Summary

Bad mobile strategies often start with bad assumptions. No one in mobile claims to know everything about mobile, it is simply too large of an ecosystem to keep tabs on all the facets. I've found the safe bet is to start with an understanding of your users' needs and to follow the seven rules covered in this chapter:

1. Forget what you think you know.
2. Believe what you see, not what you read.
3. Constraints never come first.
4. Focus on context, goals, and needs.
5. You can't support everything.
6. Don't convert, create.
7. Keep it simple.

## CHAPTER 6

Types of Mobile Applications

As we've already discussed, the mobile ecosystem is a large and deep pool. In fact, it probably isn't a pool, but an ocean—a really, really big ocean. An ocean is a good metaphor to put the different types of mobile applications in context. You see, in order to traverse an ocean, you need a sturdy boat. Boats of course come in all sorts of shapes and sizes, each with their pros and cons. Some are fast and agile, but carry little cargo. Others are large and lumbering, but can carry tons of people or cargo.

Mobile applications aren't that much different from boats in this seafaring example. You have a number of choices in what medium you use to address your goals, each with their own pros and cons. Some are quick to create but accessible to fewer users. Others address a larger market, but are far more complex and costly.

Alas, deciding what medium type (or types in some cases) to use gets only you halfway to your destination. You have to decide in what type of application context you want to present your content or information. In other words, what type of application is best suited to your problem or need? Should your app be focused on presenting information? If so, how concisely should you present it? Or is your app better suited to be an immersive experience? If yes, how immersive should it be, and is it a widget or a game? You might think it sounds silly to mention making a game, but for the mobile user, a game can be a great way to get your point across.

Although I'm a big proponent of looking at the mobile web first in the majority of cases, I find that each time I talk to a new company that wants to start with a mobile strategy, the company is simply confused about where to start. There isn't an understanding of the differences between the mobile web and a downloadable application, not to mention the pros and cons of each. Even within the mobile community, you can't tell the difference between a mobile web application and a mobile widget.

This chapter discusses the various types of mobile applications and defines each of these options. First, mobile options and their pros and cons are discussed. Those options are then discussed in an application context—how the application is presented to the user and how to leverage it effectively.

Mobile Application Medium Types

The *mobile medium type* is the type of application framework or mobile technology that presents content or information to the user. It is a technical approach regarding which type of medium to use; this decision is determined by the impact it will have on the user experience. The technical capabilities and capacity of the publisher also factor into which approach to take.

Earlier I discussed the common mobile platforms in terms of how they factor in the larger mobile ecosystem. Now we will look deeper into each of these platforms from a more tactical perspective, unpacking them, so to speak, to see what is inside.

Figure 6-1 illustrates the spectrum of mobile media; it starts with the basic text-based experiences and moves on to the more immersive experiences.

Figure 6-1. Multiple mobile application medium types

SMS

The most basic mobile application you can create is an SMS application. Although it might seem odd to consider text messages applications, they are nonetheless a designed experience. Given the ubiquity of devices that support SMS, these applications can be useful tools when integrated with other mobile application types.

Typically, the user sends a single keyword to a five-digit short code in order to return information or a link to premium content. For example, sending the keyword "freebie" to a hypothetical short code "12345" might return a text message with a coupon code that could be redeemed at a retail location, or it could include a link to a free ringtone.

SMS applications can be both "free," meaning that there is no additional charge beyond the text message fees an operator charges, or "premium," meaning that you are charged an additional fee in exchange for access to premium content.

The most common uses of SMS applications are mobile content, such ringtones and images, and to interact with actual goods and services. Some vending machines can dispense beverages when you send them an SMS; SMS messages can also be used to purchase time at a parking meter or pay lot.

A great example of how SMS adds incredible value would be Twitter, where users can receive SMS alerts from their friends and post to their timeline from any mobile device, or the SMS-to-Billboard that BBC World News put up in Midtown Manhattan (Figure 6-2).

Figure 6-2. An SMS application to interact with a billboard in Manhattan

Pros

The pros of SMS applications include:

- They work on any mobile device nearly instantaneously.
- They're useful for sending timely alerts to the user.
- They can be incorporated into any web or mobile application.
- They can be simple to set up and manage.

Cons

The cons of SMS applications include:

- They're limited to 160 characters.
- They provide a limited text-based experience.
- They can be very expensive.

Mobile Websites

As you might expect, a mobile website is a website designed specifically for mobile devices, not to be confused with viewing a site made for desktop browsers on a mobile browser. Mobile websites are characterized by their simple "drill-down" architecture, or the simple presentation of navigation links that take you to a page a level deeper, as shown in Figure 6-3.

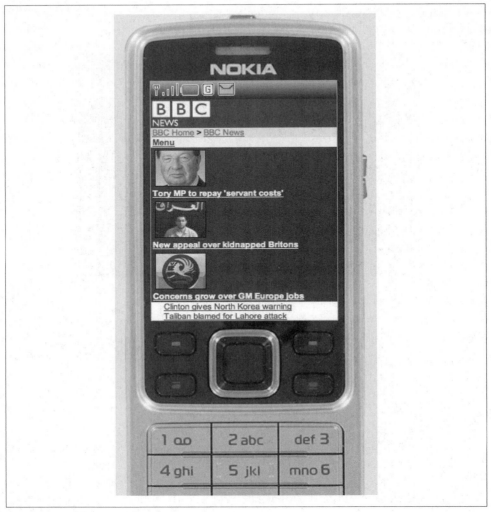

Figure 6-3. An example of a mobile website

Mobile websites often have a simple design and are typically informational in nature, offering few—if any—of the interactive elements you might expect from a desktop site. Mobile websites have made up the majority of what we consider the mobile web for the past decade, starting with the early WML-based sites (not much more than a list of links) and moving to today's websites, with a richer experience that more closely resembles the visual aesthetic users have come to expect with web content.

Though mobile websites are fairly easy to create, they fail to display consistently across multiple mobile browsers—a trait common to all mobile web mediums. The mobile web has been gradually increasing in usage over the years in most major markets, but the limited experience offered little incentive to the user. Many compare the mobile web to a 10-year-old version of the Web: slow, expensive to use, and not much to look at.

As better mobile browsers started being introduced to device platforms like the iPhone and Android, the quality of mobile websites began to improve dramatically, and with it, usage improved. For example, in just one year, the U.S. market went from being just barely in the top five consumers of the mobile web to number one, largely due to the impact of the iPhone alone.

Pros

The pros of mobile websites are:

- They are easy to create, maintain, and publish.
- They can use all the same tools and techniques you might already use for desktop sites.
- Nearly all mobile devices can view mobile websites.

Cons

The cons of mobile websites are:

- They can be difficult to support across multiple devices.
- They offer users a limited experience.
- Most mobile websites are simply desktop content reformatted for mobile devices.
- They can load pages slowly, due to network latency.

Mobile Web Widgets

Largely in response to the poor experience provided by the mobile web over the years, there has been a growing movement to establish mobile widget frameworks and platforms. For years the mobile web user experience was severely underutilized and failed to gain traction in the market, so several operators, device makers, and publishers began creating widget platforms (Figure 6-4) to counter the mobile web's weaknesses.

Trying to define what exactly a mobile web widget is and how it is different from the other mobile web media is a question for the ages. I initially saw mobile web widgets as another attempt by the mobile industry to hype a technology that no one wants. I liked to quiz mobile web widget advocates about what makes mobile web widgets different than what we can do with the mobile web. I was never able to get a straight answer. So in order to define a mobile web widget, I followed some advice from my dad: "When in doubt, look it up in the dictionary." Here was the answer I found in *Webster's Dictionary*:

> A component of a user interface that operates in a particular way.

The ever-trusty Wikipedia defines a web widget this way:

> A portable chunk of code that can be installed and executed within any separate HTML-based web page by an end user without requiring additional compilation.

Between these two definitions is a better answer:

> A mobile web widget is a standalone chunk of HTML-based code that is executed by the end user in a particular way.

Figure 6-4. An example mobile web widget

Basically, mobile web widgets are small web applications that can't run by themselves; they need to be executed on top of something else. I think one reason for all the confusion around what is a mobile web widget is that this definition can also encompass any web application that runs in a browser. Opera Widgets, Nokia Web RunTime (WRT), Yahoo! Blueprint, and Adobe Flash Lite are all examples of widget platforms that work on a number of mobile handsets. Using a basic knowledge of HTML (or vector graphics in the case of Flash), you can create compelling user experiences that tap into device features and, in many cases, can run while the device is offline.

Widgets, however, are not to be confused with the utility application context, a user experience designed around short, task-based operations, which I will explain in depth later in this chapter.

Pros

The pros of mobile web widgets are:

- They are easy to create, using basic HTML, CSS, and JavaScript knowledge.
- They can be simple to deploy across multiple handsets.
- They offer an improved user experience and a richer design, tapping into device features and offline use.

Cons

The cons of mobile web widgets are:

- They typically require a compatible widget platform to be installed on the device.
- They cannot run in any mobile web browser.
- They require learning additional proprietary, non-web-standard techniques.

Mobile Web Applications

Mobile web applications are mobile applications that do not need to be installed or compiled on the target device. Using XHTML, CSS, and JavaScript, they are able to provide an application-like experience to the end user while running in any mobile web browser. By "application-like" experience, I mean that they do not use the drill-down or page metaphors in which a click equals a refresh of the content in view. Web applications allow users to interact with content in real time, where a click or touch performs an action within the current view.

The history of how mobile web applications came to be so commonplace is interesting, and is one that I think can give us an understanding of how future mobile trends can be assessed and understood. Shortly after the explosion of Web 2.0, web applications like Facebook, Flickr, and Google Reader hit desktop browsers, and there was discussion of how to bring those same web applications to mobile devices. The Web 2.0 movement brought user-centered design principles to the desktop web, and those same principles were sorely needed in the mobile web space as well.

The challenge, as always, was device fragmentation. The mobile browsers were years behind the desktop browsers, making it nearly impossible for a mobile device to render a comparable experience. While XHTML support had become fairly commonplace across devices, the rendering of CSS2 was wildly inconsistent, and support for JavaScript, necessary or simple DHTML, and Ajax was completely nonexistent.

To make matters worse, the perceived market demand for mobile web applications was not seen as a priority with many operators and device makers. It was the classic chicken-or-the-egg scenario. What had to come first, market demand to drive browser innovation or optimized content to drive the market?

With the introduction of the first iPhone, we saw a cataclysmic change across the board. Using WebKit, the iPhone could render web applications not optimized for mobile devices as perfectly usable, including DHTML- and Ajax-powered content. Developers quickly got on board, creating mobile web applications optimized mostly for the iPhone (Figure 6-5). The combination of a high-profile device with an incredibly powerful mobile web browser and a quickly increasing catalog of nicely optimized experiences created the perfect storm the community had been waiting for.

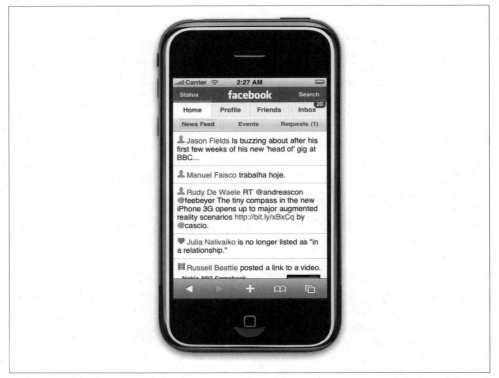

Figure 6-5. The Facebook mobile web app

Usage of the mobile web exploded with not just users of the iPhone, but users of other handsets, too. Because web applications being created for the iPhone were based on web standards, they actually worked reasonably well on other devices. Operators and device makers saw that consumers wanted not just the mobile web on their handsets, but the regular Web, too.

In less than a year, we saw a strong unilateral move by all operators and devices makers to put better mobile web browsers in their phones that could leverage this new application medium. We have not seen such rapid innovation in mobile devices since the inclusion of cameras.

The downside, of course, like all things mobile-web-related, is that not all devices support the capability to render mobile web applications consistently. However, we do see a prevalent trend that the majority of usage of the mobile web is coming from the devices with better browsers, in some markets by a factor of 7:1. So although creating a mobile web application might not reach all devices, it will reach the devices that create the majority of traffic.

Pros

The pros of mobile web applications are:

- They are easy to create, using basic HTML, CSS, and JavaScript knowledge.
- They are simple to deploy across multiple handsets.
- They offer a better user experience and a rich design, tapping into device features and offline use.
- Content is accessible on any mobile web browser.

Cons

The cons of mobile web applications are:

- The optimal experience might not be available on all handsets.
- They can be challenging (but not impossible) to support across multiple devices.
- They don't always support native application features, like offline mode, location lookup, filesystem access, camera, and so on.

Native Applications

The next mobile application medium is the oldest and the most common; it is referred to as native applications, which is actually a misnomer because a mobile web app or mobile web widget can target the native features of the device as well. These applications actually should be called "platform applications," as they have to be developed and compiled for each mobile platform.

These native or platform applications are built specifically for devices that run the platform in question. The most common of all platforms is Java ME (formerly J2ME). In theory, a device written as a Java ME MIDlet should work on the vast majority of feature phones sold around the world. The reality is that even an application written as a Java ME MIDlet still requires some adaptation and testing for each device it is deployed on.

In the smartphone space, the platform SDKs get much more specific. Although many smartphones are also powered by Java, an operating system layer and APIs added to allow developers to more easily offload complex tasks to the API instead of writing methods from scratch. In addition to Java, other smartphone programming languages include versions of C, C++, and Objective-C (Figure 6-6).

Figure 6-6. A native application in the iPhone

Creating a platform application means deciding which devices to target, having a means of testing and certification, and a method to distribute the application to users. The vast majority of platform applications are certified, sold, and distributed either through an operator portal or an app store. It is possible to create a Java ME MIDlet application and publish it for free on the Web, but it is rarely done.

Because platform applications sit on top of the platform layer, they can tap into the majority of the device features, working online or offline, accessing the location and the filesystem—and if there's camera on the device, then you can probably do something with it as well. Hence the need for certification before the application is distributed, to ensure that no one distributes an application that steals a user's personal data or maliciously uses the device to spread viruses.

However, if you exclude games, the majority (by some estimates, as much as 70 percent) of native applications in use today could be created with a little bit of XHTML, CSS, and JavaScript—in other words, a mobile web application, with little or no feature loss to the user. The advantage is that a mobile application can be developed faster, will work on more devices, require less testing, and be updated more transparently than a native application, which requires third-party certification and publishing in order to get on users' devices. All of these aspects are highly desired in the platform application space. The downside is that it requires a fast and capable mobile web browser that supports offline data and access to device features like location.

Pros

The pros of native applications include:

- They offer a best-in-class user experience, offering a rich design and tapping into device features and offline use.
- They are relatively simple to develop for a single platform.
- You can charge for applications.

Cons

The cons of native applications include:

- They cannot be easily ported to other mobile platforms.
- Developing, testing, and supporting multiple device platforms is incredibly costly.
- They require certification and distribution from a third party that you have no control over.
- They require you to share revenue with the one or more third parties.

Games

The final mobile medium is games, the most popular of all media available to mobile devices. Technically games are really just native applications that use the similar platform SDKs to create immersive experiences (Figure 6-7). But I treat them differently from native applications for two reasons: they cannot be easily duplicated with web technologies, and porting them to multiple mobile platforms is a bit easier than typical platform-based applications.

Although you can do many things with a powerful mobile web browser, creating an immersive gaming experience is not one of them—at least not yet. Seeing as how we have yet to see these types of gaming experiences appear on the desktop using standard web technologies, I believe we are still a few years out from seeing them on mobile devices. Adobe's Flash and the SVG (scalable vector graphics) standard are the only way to do it on the Web now, and will likely be how it is done on mobile devices in the future, the primary obstacle being the performance of the device in dealing with vector graphics.

The reason games are relatively easy to port ("relatively" being the key word), is that the bulk of the gaming experience is in the graphics and actually uses very little of the device APIs. The game mechanics are the only thing that needs to adapted to the various

Figure 6-7. An example game for the iPhone

platforms. Like in console gaming, there are a great number of mobile game porting shops that can quickly take a game written in one language and port it to another.

These differences, in my mind, are what make mobile games stand apart from all other application genres—their capability to be unique and difficult to duplicate in another application type, though the game itself is relatively easy to port. Looking at this model for other application areas—namely, the mobile web—could provide helpful insight into how we create the future of mobile web applications.

Pros

The pros of game applications are:

- They provide a simple and easy way to create an immersive experience.
- They can be ported to multiple devices relatively easily.

Cons

The cons of game applications are:

- They can be costly to develop as an original game title.
- They cannot easily be ported to the mobile web.

Mobile Application Media Matrix

In summary, to aid in comparing and contrasting which of these mobile application media is best for your mobile product, I've placed them into a matrix (Table 6-1).

Table 6-1. Mobile application media matrix

	Device support	Complexity	User experience	Language	Offline support	Device features
SMS	All	Simple	Limited	N/A	No	None
Mobile websites	All	Simple	Limited	HTML	No	None
Mobile web widgets	Some	Medium	Great	HTML	Limited	Limited
Mobile web applications	Some	Medium	Great	HTML, CSS, JavaScript	Limited	Limited
Native applications	All	Complex	Excellent	Various	Yes	Yes
Games	All	Complex	Excellent	Various	Yes	Yes

Application Context

Once your application medium is decided upon, it is time to look at the application context, or the appropriate type of application to present to the user in order for the user to process and understand the information presented and complete her goals. Where the application medium refers mostly to the technical approach of creating an application, the application context deals with the user experience. As discussed in Chapter 4, context is the surroundings in which information is processed, and the application user experience is no different.

Applications can be presented in a variety of ways, ranging from a simple task-based utility to an experience meant to consume the user's focus and attention. There of course is no right or wrong direction—only what is best for your user. In fact, nothing says that you can't use multiple application contexts within the same application—I just wouldn't recommend it unless you have really thought out the flow of your application, because typically it is best to present only one application context so as to avoid confusing the user. If you think it best for your app to mix contexts, then give the user the option to switch between them; for example, some smartphones allow for an orientation change, so if the device is rotated to landscape mode, your app switches from an informative view to a utility view, or maybe from a locale view to an immersive view.

Utility Context

The most basic application context is the utility, or a simple user experience metaphor that is meant to address short, task-based scenarios. Information is meant to be presented in a minimal fashion, often using the least amount of user input as possible. An example of a utility might be a calculator, weather forecast, unit conversion, stocks, world clock, and so on. In each of these cases, the user enters a small amount of information into the utility, like a simple equation, a city, or a stock symbol, and either by performing a small background task or fetching information online, the utility is able to present data to the user in the desired context (Figure 6-8).

Figure 6-8. An example utility application

The goal of the utility is to give users at-a-glance information, therefore offering users a minimal design aesthetic, focusing the design around the content in view, and often using larger type and a sparse layout.

It would be easy to mistake utilities for widgets, given that widgets are a "component of a user interface that operates in a particular way." But utilities can be much more than widgets; they are not merely an extension of the user experience, but are applications in their own right that can establish their own look and feel separate from the established user experience.

Use utilities for short, simple tasks, at-a-glance information, when there is limited content to display, and when combined with an immersive context to create dual-mode applications.

Locale Context

The locale context is a newer application type that is still being defined by the mobile community, but we are certainly seeing some clear patterns of how to create locale applications (Figure 6-9). As more location information is being published online, and more devices add GPS to pinpoint the user's location, locale is becoming an excellent data point to pivot information around. For example, I can use location to display the cafés nearest to my current location. Plus I can layer multiple data sources into the application, such as: of the cafés nearest to me, which ones have free wireless access? Or, do I have friends in the general area that can meet me?

Figure 6-9. An example locale application

Locale applications almost always have at least one thing in common: a map on which they plot the requested data points visually. At the very least, they list items in order of distance, with the nearest item first and the farthest last. The user's goal is to find information relative to his present location, and content should always be designed with this in mind. When creating locale apps, it is important to ensure that the user's present location is always clearly identified, as well as a means of adding data to it. This could be another location, in the case of finding point-to-point directions, or it could be a keyword query to find people, places, or things nearby.

Use locale applications for location-based applications, applications that might contain a dynamic map, and listing multiple location-based points of interest.

Informative Applications

The informative application is an application context in which the one and only goal is to provide information, like a news site, an online directory, a marketing site, or even a mobile commerce site, where the key task of the user is to read and understand and it is not necessary to interact (Figure 6-10). This isn't to say that you cannot include calls to action in the informative context—in fact, you should, but they should be based around what you can assume about your users in this context.

For example, remember that most mobile tasks are short and are often undertaken during brief idle periods. The user doesn't have much extra time and the task can be interrupted at any moment. In the case of a mobile news site, provide the user with the option to mark a page or story to be read later. With an online directory, allow the user to flag favorite entries. With a marketing site, allow users to enter the shortest possible contact information, like their phone number or email. And with a mobile commerce site, allow users to save items to a wishlist to review and purchase later.

The theme here is that although reading information is a simple task, it usually creates a complex chain of events that can be anticipated. With mobile applications, we want to avoid forcing the users to input too much information with their mobile devices, which is more difficult and takes more time than it would on another medium such as a desktop or laptop computers. Instead, we want to look for ways we can interconnect experiences, having users use the informative context to filter to the most desirable information when they have a moment, and allowing them to interact with it later, when they have more time, from the medium of their choice.

Use informative applications when users need information to gain an understanding or perform a physical task, for applications that can be used in multimedia contexts such as desktop and mobile, for information-heavy applications, and for marketing or promotional applications.

Figure 6-10. An example informative application

Productivity Application Context

The productivity application context is used for content and services that are heavily task-based and meant to increase the users' sense of efficiency. With these types of applications, we can assume that the users are more committed to accomplishing a particular goal, like managing content such as messages, contacts, or media, but we should still assume that they are doing so during idle periods (Figure 6-11). Just because the application context is meant to make users more productive, we can't assume that they are able to make the same time commitment as they would in the desktop context.

Productivity applications are often very structured, presenting information in a defined hierarchy and often using the folder or group metaphor to define a sense of order to the user. When designing these types of apps, we need to pay careful consideration to how the user thinks out the task. People have an uncanny ability to understand and recall complex hierarchies of tasks—for example, what they need to do first, second, and third in order for a particular solution to work. We take this for granted and in the desktop context often show the users the entire hierarchy visually. In the mobile context, we don't have the screen real estate, and therefore need to help users find their way.

Figure 6-11. An example productivity application

One method is to focus on prioritization of tasks; productivity applications typically include some method of direct or indirect prioritization. If we look at a mobile email client, we see that the app generally focuses around the inbox, which is the top-priority item, given that all new messages will come there first. All other folders are of a lower priority, as in order for messages to get there, we will have had to process them previously. We can use this screen as a central focus point, assuming that users will spend the majority of their time there, and branching out onto other screens from this central spot. But we can't forget other high-priority items, like the ability to send a new message or create a new item. This is a task that is typically included on every screen within an email app, and in the same position throughout to ensure that users always have quick access to create a new item.

The productivity context is one of the hardest application contexts to get right, so do yourself a favor and start simple. Start with one feature, treat it almost as if it were a single focus utility, and get it right before you move on to the next. Layer in your features one at a time until you feel like you have met the users' goals, and stop the moment that it becomes it bit overwhelming to manage. You probably won't be able to include every feature, so you will need to include only the ones that are most important to users, and lose everything else.

Use the productivity application context for information-heavy applications where the user will need to manage content from a mobile device and for heavily structured, hierarchy-based tasks.

Immersive Full-Screen Applications

The final application context is an immersive full-screen application, like a game, a media player, or possibly even a single-screen utility. These applications are meant to consume the user's focus, often doing so by filling the entire screen (Figure 6-12), and leaving no trace of the device user interface to distract the user. Again, the majority of mobile engagement occurs when the user has idle periods of time; the immersive context is typical in most entertainment applications, one of the most popular mobile content areas.

Figure 6-12. An example of an immersive application

The most common use of the immersive context is obviously with a game, for which you want the user to focus on how to play the game. But this context can also be used with other contexts, presenting a full-screen view of content when the device orientation changes in many higher-end devices. For example, if we were making a locale-based application, we could add a feature that changes the user experience to the immersive context, showing a full-screen map, or point-by-point directions, whenever the device is held in landscape view. This is not a feature that many applications include, but I think it is worth considering.

Even with mobile web apps, many devices allow for detection of an orientation change. Typically, the app just scales to fill the page, actually breaking the intended user experience, but by adding the orientation-specific styles, the designer could create an immersive version of the application, presenting the app content in a more at-a-glance, friendly way, helpful for devices placed on automobile dashboards, or held in the hand to show others.

Use an immersive full-screen application for games, media players, and alternative views of another application context.

Application Context Matrix

I put each of the application contexts into Table 6-2, comparing and contrasting their benefits to help you determine what is best for your application.

Table 6-2. Application context matrix

	User experience type	Task type	Task duration	Combine with
Utility	At-a-glance	Information recall	Very short	Immersive
Locale	Location-based	Contextual information	Quick	Immersive
Informative	Content-based	Seek information	Quick	Locale
Productivity	Task-based	Content management	Long	Utility
Immersive	Full screen	Entertainment	Long	Utility, locale

As you can see, mobile applications can run the gamut from intense experiences to simple tools. In some cases, they can switch back and forth between the two. Figure out which type of application is best for your users and in what context.

## CHAPTER 7

Mobile Information Architecture

Your information architecture (also known as IA), is the foundation of your mobile product. A well-engineered product with good visual design can still fail because of poor information architecture. The truly successful mobile products always have a well-thought-out information architecture.

From a simple mobile website to an iPhone application, the mobile information architecture defines not just how your information will be structured, but also how people will interact with it. This is made especially tricky when you consider that different devices have different capabilities and therefore different interaction models. Take the way people interact with their devices: for example, a touch device on which the user literally points and clicks, or a more basic device on which the user uses the directional pad to navigate to the desired location.

Also, don't forget that the information architecture emphasizes how you address context. In other words, a good mobile information architecture is based around the various user contexts that I talked about in Chapter 4. The secret is that mobile information architecture isn't all that different from how you might architect software or a website; it just has a few added challenges.

What Is Information Architecture?

Before we get into the specifics of mobile information architecture, let's first talk about exactly what information is. I can think of no better definition than the seminal O'Reilly book *Information Architecture for the World Wide Web* (*http://oreilly.com/catalog/9780596527341/*) by Morville and Rosenfeld, otherwise known in information architect circles as the "polar bear" book. This definition outlines the following:

- The structural design of shared information environments
- The combination of organizations, labeling, search, and navigation systems within websites and intranets

- The art and science of shaping information products and experiences to support usability and findability
- An emerging discipline and community of practice focused on bringing principles of design and architecture to the digital landscape

Similar to how mobile technology has many facets, so does information architecture, as it is often used as an umbrella term to describe several unique disciplines, including the following:

Information architecture
> The organization of data within an informational space. In other words, how the user will get to information or perform tasks within a website or application.

Interaction design
> The design of how the user can participate with the information present, either in a direct or indirect way, meaning how the user will interact with the website of application to create a more meaningful experience and accomplish her goals.

Information design
> The visual layout of information or how the user will assess meaning and direction given the information presented to him.

Navigation design
> The words used to describe information spaces; the labels or triggers used to tell the users what something is and to establish the expectation of what they will find.

Interface design
> The design of the visual paradigms used to create action or understanding.

In 2000, information architect Jesse James Garrett created the Elements of User Experience (*http://www.jjg.net/elements/pdf/elements.pdf*, also seen in Figure 7-1),[*] to help show how each of the layers of the user experience come together to create both hypertext experiences and the more software-like application experience.

As you can see, information architecture composes the core of the user experience. The role of information architecture is played by a variety of people, from product managers to designers and even developers. To make things more confusing, information architecture can be called many different things throughout the design and development process. Words like intuitive, simple, findable, usable, or the executive favorite—easy-to-use—all describe the role that information architects play in creating digital experiences.

The visual design of your product, what frameworks you use, and how it is developed are integral to the success of any product, but the information architecture stands apart as being the most crucial element of your product. It is the first line of scrimmage—the user's first impression of your product. Even if you have the best design, the best code,

---

[*] Image used with permission of Jesse James Garrett (*http://jjg.net/elements/*).

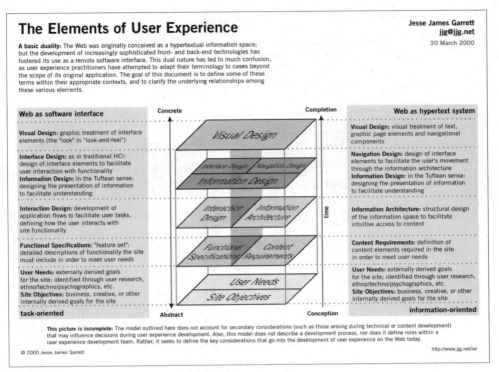

Figure 7-1. Jesse James Garrett's Elements of User Experience

and the best backend service, if the user cannot figure out how to use it, she will fail—and so will your product.

Mobile Information Architecture

Although information architecture has become a common discipline in the web industry, unfortunately, the mobile industry—like software—has only a handful of specialized mobile information architects. Although mobile information architecture is hardly a discipline in its own right, it certainly ought to be. This is not because it is so dissimilar from its desktop cousin, but because of context, added technical constraints, and needing to display on a smaller screen as much information as we would on a desktop.

For example, if we look at the front page of *http://www.nytimes.com* as seen from a desktop web browser compared to how it may render in a mobile browser (Figure 7-2), we see a content-heavy site that works well on the desktop, and is designed to present the maximum amount of information above the "fold" or where the screen cuts off the content. However, in the mobile browser, the text is far too small to be useful.

Figure 7-2. Comparing the New York Times website in desktop and mobile browsers

The role of a mobile information architect would be to interpret this content to the mobile context. Do you use the same structure, or sections? Do you present the same information above the fold? If so, how should that be prioritized? How does the user navigate to other areas? Do you use the same visual and interaction paradigms, or invent new ones? And if you do start to invent new paradigms, will you lose the visual characteristics of what users expect?

These are only some of the questions you have to ask yourself when starting to create a mobile information architecture. As you can see in Figure 7-3 there are several different ways that the *New York Times* has been interpreted for the mobile context.

Figure 7-3. The many mobile experiences of the New York Times

But we are only beginning to scratch the surface. We also need to design our mobile information architecture to address the mobile context. Given that many devices can detect our current location, which is one of the most immediate types of context, how does the *New York Times* application address the user's context? For example, as a publication that serves both New York City and a larger global audience, if I'm not in New York, should I still see the local New York headlines? Or should I see the headlines based on my location?

Keeping It Simple

When thinking about your mobile information architecture, you want to keep it as simple as possible.

Support your defined goals

If something doesn't support the defined goals, lose it. Go back to your user goals and needs, and identify the tasks that map to them. Find those needs and fill them.

Ask yourself: what need does my application fill? What are people trying to do here? What is their primary goal? Once you understand that, it is a simple process of reverse-engineering the path from where they want to be to where they are starting. Cut out everything else—your site or application doesn't need it. For example, to get some news and information on a mobile device, you need to first ask what the goal is. What is the need you are trying to fill? Then you need to apply context. Where are your users? What are they doing? Are they waiting for the bus? Do they have only a minute to spare? Or, do they have five minutes to spare? With these answers, you get your information architecture.

Clear, simple labels

Good trigger labels, the words we use to describe each link or action, are crucial in Mobile. Words like "products" or "services" aren't good trigger labels. They don't tell us anything about that content or what we can expect. Now, I would argue that good trigger labels are crucial in the Web as well, that we've become lazy and we assume so much about the user that we ignore the use of good trigger labels. Users have a much higher threshold of pain when clicking about on a desktop site or application, hunting and pecking for tasty morsels. Mobile performs short, to-the-point, get-it-quick, and get-out types of tasks. What is convenient on the desktop might be a deal breaker on mobile.

Keep all your labels short and descriptive, and never try to be clever with the words you use to evoke action. The worst sin is to introduce branding or marketing into your information architecture; this will just serve to confuse and distract your users. Executives love to use the words they use internally to external communications on websites and applications, but these words have no meaning outside of your company walls. If the user is just trying to get music, don't call it "My Music," "My MP3s," or something made up that only strokes our corporate egos, such as "AudioJams™"—just call it "Music."

Don't try to differentiate your product offering by what you call it. Create something unique by creating a usable and intuitive experience based on focusing on what users need and using the same language they use to describe those needs.

Based on what we know from web design, you should use simple, direct terms for navigating around your pages rather than overly clever terms. That latter typically result in confused visitors who struggle to find the content they are looking for. When that happens, they will go elsewhere to look for the information they want. So, if you apply these same mistakes to a constrained device like mobile, then you end up adding confusion to the user experience at a higher magnitude than the Web.

Site Maps

The first deliverable we use to define mobile information architecture is the site map. Site maps are a classic information architecture deliverable. They visually represent the relationship of content to other content and provide a map for how the user will travel through the informational space, as shown in Figure 7-4.

Mobile site maps aren't that dissimilar from site maps we might use on the Web. But there are a few tips specific to mobile that we want to consider.

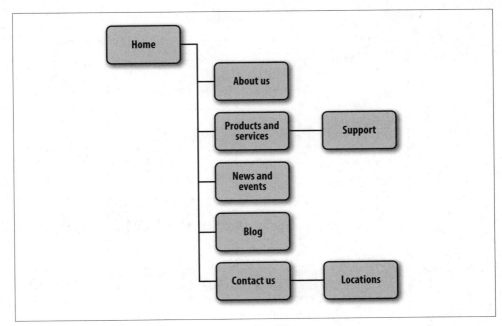

Figure 7-4. An example mobile site map

Limit opportunities for mistakes

Imagine a road with a fork in it. We can go either left or right. The risk that we will make the wrong choice is only 50 percent, meaning that we have a better than good chance that we will get to where we want to go. But imagine three roads. Now our chances have dropped to 33 percent. Four roads drops your chances to 25 percent, and five roads takes you down to 20 percent. Now a 20 percent chance isn't great, but it isn't too bad, either.

Now think of your own website. How many primary navigation areas do you have? Seven? Eight? Ten? Fifteen? What risk is there to the users for making a wrong choice? If they go down the wrong path, they can immediately click back to where they started and go down another path, eliminating the wrong choices to find the right ones. The risks for making the wrong choice are minor.

In Figure 7-5, you can see a poorly designed mobile information architecture that too closely mimics its desktop cousin; it was not designed with the mobile user in mind.

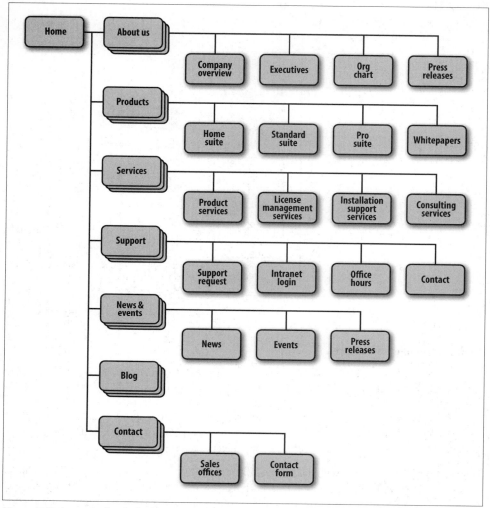

Figure 7-5. An example of a bad mobile information architecture that was designed with desktop users in mind rather than mobile users

But in mobile, we cannot make this assumption. In the mobile context, tasks are short and users have limited time to perform them. And with mobile websites, we can't assume that the users have access to a reliable broadband connection that allows them to quickly go back to the previous page. In addition, the users more often than not have to pay for each page view in data charges. So not only do they pay cash for viewing the wrong page by mistake, they pay to again download the page they started from: we can't assume that pages will be cached properly.

Therefore, my advice is to limit users' options, those forks in the road, to five or less. Anything more, and you introduce far too much risk that the user will make a mistake and head off in the wrong direction.

Confirm the path by teasing content

After the users have selected a path, it isn't always clear whether they are getting to where they need to be. Information-heavy sites and applications often employ nested or drill-down architectures, forcing the user to select category after category to get to their target. To reduce risking the user's time and money, we want to make sure we present enough information for the user to wade through our information architecture successfully. On the Web, we take these risks very lightly, but with mobile, we must give our users a helping hand. We do this by teasing content within each category—that is, providing at least one content item per category.

I've done a fair share of mobile sites and applications that sell ringtones, the historical mobile content motherlode. The challenge with ringtone sites is you have a lot of items, grouped by artist, album, genre, and so on. The user starts with a goal like "I want a new ringtone" and finds an item that suits his taste within a catalog of tens of thousands of items.

In order to make sense of a vast inventory of content, we have to group, subgroup, and sometimes even subgroup again, creating a drill-down path for the user to browse. Though on paper this might seem like a decent solution, once you populate an application with content, the dreaded "Page 1 of 157" appears. What user would ever sit there with a mobile device and page through 157 pages of ringtones? What user would page through five pages of content?

On an early site I worked on, users would flip through a few pages of content, then give up or go back and visit another area. We could see a direct relationship to the number of pages viewed to sales—essentially, more pages loaded meant fewer sales. Then we realized we could take the most popular item based on sales and place it as the first item in any list, which is teasing the content.

In Figure 7-6, you can see in a constrained screen that teasing the first few items of the page provides the user with a much more intuitive interface, immediately indicating what type of content the user can expect.

We immediately saw that users were finding content more quickly, driving up our sales. It was like night and day. Since those days, I've tested this principle on a variety of mobile sites—not just ringtones, but game catalogs, news sites, and regular old corporate websites. Each time, it has improved the conversion, getting users to the content they seek with the least amount of backtracking.

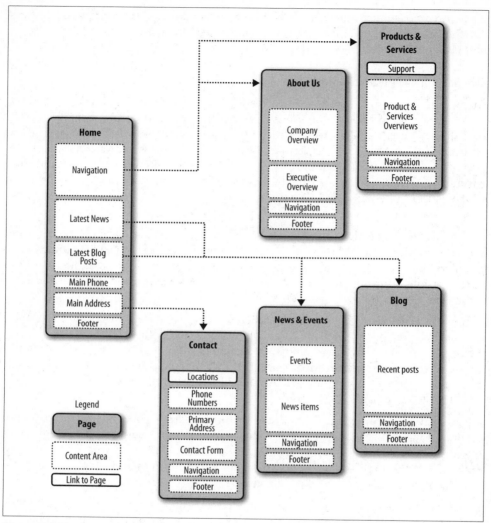

Figure 7-6. Teasing content to confirm the user's expectations of the content within

Clickstreams

Clickstream is a term used for showing the behavior on websites, displaying the order in which users travel through a site's information architecture, usually based on data gathered from server logs. Clickstreams are usually historical, used to see the flaws in your information architecture, typically using heat-mapping or simple percentages to show where your users are going. I've always found them to be a useful tool for rearchitecting large websites.

However, information architecture in mobile is more like software than it is the Web, meaning that you create clickstreams in the beginning, not the end. This maps the ideal path the user will take to perform common tasks. Being able to visually lay out the path users will take gives you a holistic or bird's-eye view of your mobile information architecture, just as a road map does. When you can see all the paths next to each other and take a step back, you start to see shortcuts and how you can get users to their goal faster or easier, as shown in Figure 7-7.

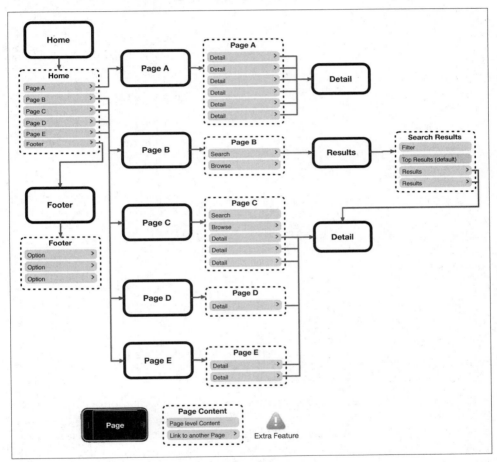

Figure 7-7. An example clickstream for an iPhone web application

Now the business analyst in you is probably saying, "Just create user or process flows," such as the esoteric diagram shown in Figure 7-8, which is made up of boxes and diamonds that look more like circuit board diagrams than an information architecture.

If that is what your team prefers, then by all means, flow away. Personally, I like to present all of my information architecture deliverables from the perspective of the user,

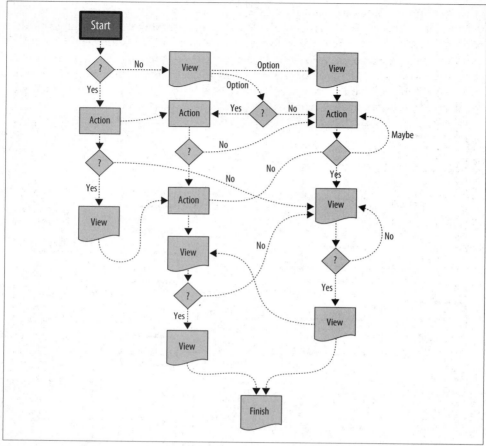

Figure 7-8. An example process flow diagram

using the same metaphors she will use to make her way through my information architecture—in this case, either a screen or page metaphor.

A good architect's job is to create a map of user goals, not map out every technical contingency or edge case. Too often, process flows go down a slippery slope of adding every project requirement, bogging down the user experience with unnecessary distractions, rather than focusing on streamlining the experience. Remember, in mobile, our job is to keep it as simple as possible. We need to have an unwavering focus on defining an excellent user experience first and foremost. Anything that distracts us from that goal is just a distraction.

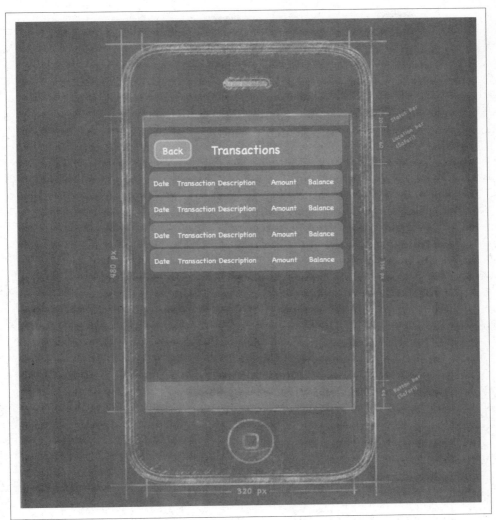

Figure 7-9. An example of an iPhone web application wireframe, intended to be low fidelity to prevent confusion of visual design concepts with information design concepts

Wireframes

The next information architecture tool at our disposal is wireframes. *Wireframes* are a way to lay out information on the page, also referred to as *information design*. Site maps show how our content is organized in our informational space; wireframes show how the user will directly interact with it. Wireframes are like the peanut butter to the site map jelly in our information architecture sandwich. It's the stuff that sticks. Wireframes like the one in Figure 7-9 serve to make our information space tangible and useful.

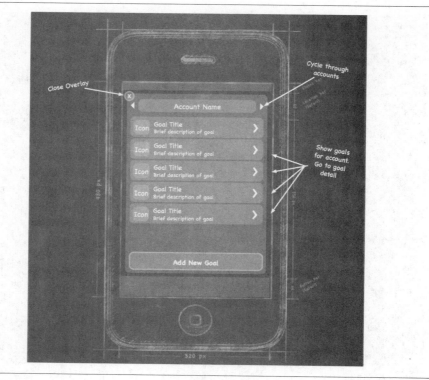

Figure 7-10. Using annotations to indicate the desired interactions of the site or application

But the purpose of wireframes is not just to provide a visual for our site map; they also serve to separate layout from visual design, defining how the user will interact with the experience. How do we lay out our navigation? What visual or interaction metaphors will we use to evoke action? What are the best ways to communicate and show information in the assumed context of the user? These questions and many more are answered with wireframes.

Although I've found wireframes to be one of the most valuable information deliverables to communicate my vision for how a site or app will work, the challenge is that a diagram on a piece of paper doesn't go a long way toward describing how the interactions will work. Most common are what I call "in-place" interactions, or areas where the user can interact with an element without leaving the page. This can be done with Ajax or a little show/hide JavaScript. These interactions can include copious amounts of annotation, describing each content area in as much length as you can fit in the margins of the page, as shown in Figure 7-10.

At this point, I highly recommend that you get some feedback from either others on your project or my most trusted reviewer, my wife. Well, not *my* wife, but someone you know and trust—and the less technical, the better. Have her review your work as you iterate through ideas. Describe what problems you are trying to solve and ask her what she is thinking. It has been an invaluable process for me over the years, not specifically for the feedback I receive—though you do get the occasional strokes of genius—but for forcing me to communicate what I've done and verbalize my thoughts. You'd be surprised how often you get an idea in your head that you think is brilliant, but once you say it out loud, it just sounds absurd.

In mobile, it is this kind of feedback, using wireframes as your key deliverable, that turns good ideas into excellent mobile products.

Prototyping

As mentioned before, wireframes lack the capability to communicate more complex, often in-place, interactions of mobile experiences. This is where prototypes come in. Prototypes might sound like a scary (or costly) step in the process. Some view them as redundant or too time-consuming, preferring to jump in and start coding things. But as with wireframes, I've found that each product we've built out some sort of prototype has saved both time and money. The following sections discuss some ways to do some simple and fast mobile prototyping.

Paper prototypes

The most basic level we have is paper prototyping: taking our printed-out wireframes or even drawings of our interface, like the one shown in Figure 7-11, and putting them in front of people.

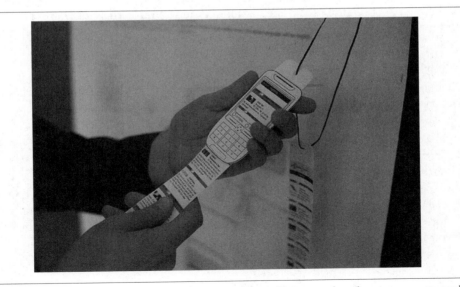

Figure 7-11. A paper prototype, where the interaction is nothing more than drawings on note cards

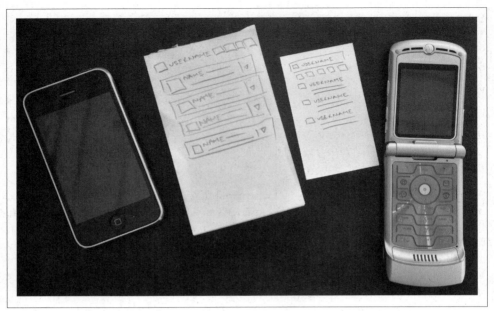

Figure 7-12. A touch interface paper prototype next to its smaller sibling

Create a basic script of tasks (hopefully based on either context or user input) and ask users to perform them, pointing to what they would do. You act as the device, changing the screens for them. For paper prototypes, I don't recommend using full sheets of paper; instead try using small blank note cards, and for lower-end devices, use business-card-sized paper (Figure 7-12). The size matters and you'll learn as much from how the user manages working with small media as you will what information is actually on it.

Context prototype

The next step is creating a context prototype (Figure 7-13). Take a higher-end device that enables you to load full-screen images on it. Take your wireframes or sketches and load them onto the device, sized to fill the device screen. Leave the office. Go for a walk down to your nearest café. Or get on a bus or a train. As you are traveling about, pull out your device and start looking your interface in the various contexts you find yourself currently in.

Pay particular attention to what you are thinking and your physical behavior while you are using your interface and then write it down. If you are brave and don't have strict nondisclosure issues, ask the people around you to use it, too. I wouldn't bother with timing interactions or sessions, but try to keep an eye on a clock to determine how long the average session is.

Figure 7-13. An example of a context prototype, or taking images loaded onto a device and testing them in the mobile context

HTML prototypes

The third step is creating a lightweight, semifunctional static prototype using XHTML, CSS, and JavaScript, if available. This is a prototype that you can actually load onto a device and produce the nearest experience to the final product, but with static dummy content and data (Figure 7-14). It takes a little extra time, but it is worth the effort.

With a static XHTML prototype, you use all the device metaphors of navigation, you see how much content will really be displayed on screen (it is always less than you expect), and you have to deal with slow load times and network latency. In short, you will feel the same pains your user will go through.

Whatever route you wish to take, building a mobile prototype takes you one very big leap forward to creating a better mobile experience. I won't lie: it can feel like a thankless exercise at times, but your users will thank you in the long run by using your app.

Different Information Architecture for Different Devices

Depending on which devices you need to support, mobile wireframes can range from the very basic to the complex. On the higher-end devices with larger screens, we might be inclined to add more interactions, buttons, and other clutter to the screen, but this would be a mistake. Just because the user might have a more advanced phone, that

Figure 7-14. An XHTML prototype that you can actually interact with on real mobile devices

doesn't mean that he is giving you license to pack his screen with as much information as you can muster.

The motivations, goals, and how users will interact with a mobile experience are the same at the low end as they are on a high-end device. For the latter, you just have better tools to express the content. You can learn a lot from designing for the lower end first. The greatest challenge in creating valuable experiences is knowing when to lose what you don't need. You don't have a choice on lower-end devices—it must be simple. When designing for both, it is best to try and to keep your information architecture as close to each other as possible without sacrificing the user experience. They say that simple design is the hardest design, and this principle certainly is true when designing information architecture for mobile devices.

The Design Myth

A little secret about interactive design is that people don't respond to the visual aesthetic as much as you might think. What colors you use, whether you use square or rounded corners, or, gradients or flat backgrounds, helps build first impressions, but it doesn't do too much to improve the user's experience. Don't get me wrong: users appreciate good design, but they are quickly indifferent about the visual aesthetic and move almost immediately to the layout (information design), what things are called (taxonomy), the findability of content, and how intuitive it is to perform tasks. These are all facets of information architecture.

Just look at one of the top-selling iPhone Twitter applications, Tweetie, shown in Figure 7-15. Many consider Tweetie to be a "well-designed" application, but because it is built from the same API as all other iPhone applications, at first glance there is little that is actually visually distinctive between this and other applications. What makes this application "well designed" is how the content is applied to the context of the user—in other words, the mobile information architecture. In this example, the information design uses common layout metaphors, highlighted on the righthand side of Figure 7-15 to provide the user with familiar placement of common tasks, allowing the user to perform repetitive tasks common with most Twitter applications. The point is great information design is often mistaken for great visual design.

Figure 7-15. Comparing visual design to information design of the iPhone application Tweetie

Most non–information architects almost always do information architecture in some form or another; often, they don't even know they are doing it. They might do a few wireframes, or maybe a site map. Sometimes designers will jump in and incorporate information architecture deliverables directly into their designs. By not focusing on the information architecture exclusively from the start, you risk confusing your disciplines, your deliverables, and ultimately your direction. The more time you spend focusing on just your information architecture, the faster and less costly your project will be.

## CHAPTER 8

Mobile Design

When building mobile experiences, it is impossible to create a great experience without three ingredients: context, information architecture, and visual design. This chapter focuses on the latter ingredient of the recipe. The visual design of your experience is the direct representation of everything underneath; it is the first impression the user will have. A great design gives the user high expectations of your site or application; a poor design leads to lower expectations.

Users' expectations translate to value and trust. In the mobile space, where content is often "free" (they still need to pay for data charges), users often have low expectations, due to the limitations of the canvas. Confusing site structures and slow download speeds reinforce those initial low expectations. In the downloadable application space, where application design can be much more robust, users must purchase applications almost sight unseen. For example, they may see just a small screenshot of your application or game. But if the application doesn't meet the higher expectations of the design, your application downloads will drop like a stone. The design, that first impression, determines right from the start if the user will spend five seconds, five minutes, or five hours with your product.

This leads us to the most significant challenge in mobile design: creativity. You can't always be as creative as you want to be. Many devices just can't support complex designs for every channel; for example, on many lower-end devices, the mobile web experience may just be a list of links. But every device has the capability to create a best-in-device experience; it just depends on how you take advantage of the application medium and context that you plan to use.

On computers, there is a strategy called "lowest common denominator": in order to reach the widest possible number of platforms, you create a product that works on the most common architectural components on all platforms (see Figure 8-1). Well, in computers, where you may have under a dozen different platforms, this is a great concept, but in mobile development, where you might be dealing with hundreds of different devices, it becomes a necessity.

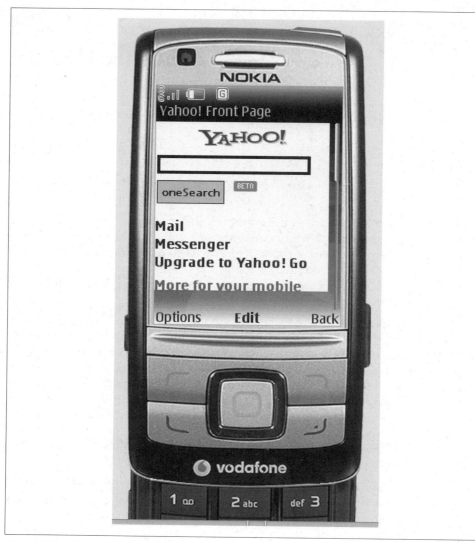

Figure 8-1. A lowest-common-denominator design

Typically, mobile design starts with the lowest common denominator. As a designer, you ask yourself, "How do I visually express this content across the most possible devices?" You start with the most basic of designs, catering to the limitations of the device. You try to pepper in some nice-looking elements until you've reached the extent that the device platform can tolerate. You are left with a Frankenstein-like design that only your mother could love.

Interpreting Design

Mobile design reminds me of a 25-year veteran graphic designer friend of mine. His days were spent creating print designs and advertisements, defining precisely what each design would be, down to the picas and points. His method of design meant creating a vision for how to communicate information or ideas in his head, and then duplicating that on the printed page. In his mind, it wasn't right unless it was exactly like his original vision.

I offered to help him with his website; looking back on this, I realize that it probably was not the wisest decision. He spent months obsessing about exactly how his site looked. I would try to explain to him that the Web isn't about creating precise experiences, like you do in print. I explained that my method of communicating information was to structure it with design, use design elements to enhance the information, and enable the user to interpret it. It is important to remember that every experience is unique. That experience depends on the user's screen size, web browser, text settings, the speed of his computer, and his connection to the Internet. There are simply too many variables for us to try to "control" the design completely.

This frustrated the veteran graphic designer. He could not look at my interpreted designs without trying to precisely position every element on the page. And if I resized the browser window, he would get angry.

I frequently think of this project when working with mobile designs. It isn't that either one of us was right or wrong. That is the great thing about design—it is completely subjective, giving designers plenty of things to argue about. We just had two very different ways of using design to express information, based on the fact that we came from different media. He wanted it to treat the design precisely, to recreate his vision exactly. I wanted it to be flexible, catering to the unknown variables of the medium. The reality is that we probably should have met someplace in the middle.

Mobile design isn't that different. Precise designs might look better, but they can be brutal to implement. More flexible designs might not be much to look at, but they work for the most users, or the lowest common denominator. But more than that, our backgrounds and our training can actually get in the way of creating the best design for the medium. We like to apply the same rules to whatever the problem in front of us might be. In mobile design, you interpret what you know about good design and translate it to this new medium that is both technologically precise and at times incredibly unforgiving, and you provide the design with the flexibility to present information the way you envision on a number of different devices.

In the end, the graphic designer and I scrapped the work, and he provided me his pica-perfect designs as giant images, which I turned into a series of massive image maps. To this day, I do not work with graphic designers on web or mobile projects.

The Mobile Design Tent-Pole

In Hollywood, executives like to use the term "tent-pole" to describe their movies and television shows. The term has dual meanings: one is business, and the other creative. The business goal of a tent-pole production is to support or prop up the losses from other productions. However, to create a tent-pole production, the creators involved must make an artistic work that they know will appeal to the largest possible audience, providing something for everyone. You probably know tent-pole movies as "blockbusters"; in television, they are known as "anchor" shows.

Trying to reach the widest possible audience poses a problem. Hollywood is learning with great pains that with so many entertainment options and with today's audience being so hard to reach through traditional advertising channels, tent-pole productions are failing. As the number of social niches increases, it becomes difficult to satisfy the specific tastes of each social group. What one group finds hysterically funny, several other groups might find offensive. Today, tent-pole productions often come off as bland and half-hearted, failing to appease anyone.

One of the most interesting examples of how the tide turned in entertainment is with the animated films of Disney versus those of Pixar. For years, Disney produced tent-pole family fare quite successfully. But as competition increased, notably from Pixar, Disney films would spend millions to create stale and dated films, losing audiences and revenue. Meanwhile, Pixar found that their movies could be successful by avoiding the traditional storytelling formats of animated film, which Disney essentially defined. Instead, Pixar based their stories around specific emotional themes and was able to connect with audiences of all ages, in multiple cultures and across multiple niches.

In 2006, Disney acquired Pixar, making its top executives the new leaders of all Disney creative projects. Although Disney technically acquired Pixar, I've always looked at it as the other way around. Disney realized that it needed to be more Pixar and less Disney in order to grow and adapt to today's changing audiences and niches. This is something that Pixar was doing correctly.

Back in the world of mobile design, the de facto strategy is to create tent-pole products. Like the old days of Disney, the strategy is to sink millions into creating tent-pole products, or products that support the largest number of devices that no one will ever use. They are creatively stale, they lack inspiration, and they simply exist with no meaningful purpose to the user. They make the same mistake Disney made, thinking that it could simply put something on the market that might not be the best quality, but because it carried the Disney name, people would buy it.

To have a successful mobile design, you have to think like Pixar. Find that emotional connection, that fundamental need that serves many audiences, many cultures, and many niches and design experiences. Too often, designers simply echo the visual trends of the day, mimicking the inspiration of others—I'm certainly guilty of it. But with mobile design, once you find that essential thing, that chewy nougat we call "context" that lives at the center of your product, then you will find ample inspiration of your own to start creating designs that translate not only across multiple devices, but across multiple media.

Sure, there are countless examples of poorly designed mobile products that are considered a success. You only need to look as far as the nearest mobile app store to find them. This is because of the sight unseen nature of mobile commerce. Traditionally, you couldn't demo—or in some cases even see—screenshots of a game or mobile application before you bought it. You simply had to purchase it and find out if it was any good.

Apple's App Store quickly changed that. You can clearly see that the best-selling games and applications for the iPhone are the ones with the best designs (Figure 8-2).

Figure 8-2. The app icon design greatly influences the user's expectation of quality

Users look at multiple screenshots (Figure 8-3), read the user reviews, and judge the product based on the quality of its icon and of the screenshots before they buy.

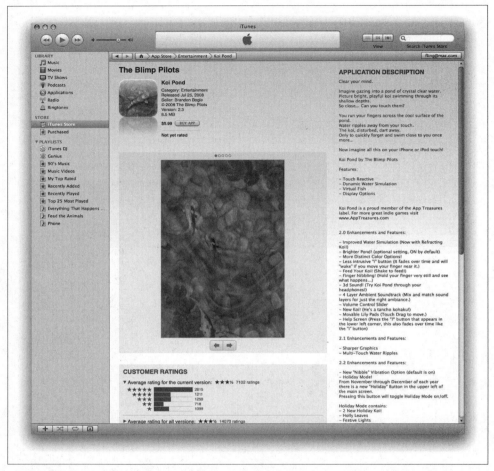

Figure 8-3. Users are able to determine the quality of the app, largely influenced by the design, before they make a purchase

The Apple App Store is proving everyday that mobile design doesn't have to start with tent-pole lowest-common-denominator products—it can instead start with providing the best possible experience and tailoring that experience to the market that wants it most.

Designing for the Best Possible Experience

When the first iPhone came out, I got in a lot of trouble from my web and mobile peers for publicly saying, "The iPhone is the only mobile device that matters right now." They would argue, "What about ABC or XYZ platforms?" My response was that those are important, but the iPhone provides the best possible experience and that is where consumers will go. Since those days, we've seen the iPhone shatter just about every record in mobile devices, becoming one of the best-selling phones ever and one of the most used mobile browsers in the world—two-thirds of mobile browsing in the U.S. comes from an iPhone or an iPod touch, not to mention that more than a billion mobile applications have been sold for these devices in under a year.

Recently, I was speaking at a conference where I ran into one of my peers, who questioned my premise that the iPhone was the most important device in mobile. He came up to me, and the first thing he said was, "I remember you telling me ages ago that the iPhone is the only device that mattered, and I didn't believe you. And here we are today focusing our business on the iPhone." It was an odd (and rare) reverse I-told-you-so moment. Here was this seasoned mobile guy telling me that his instincts had been wrong and my instincts had been right. I thought it must have been hard for him to go against his instincts and shift not just his thinking but his entire business toward supporting one popular device.

The lesson here is that although it may defy your business instincts to focus your product on just one device, in mobile development, the risks and costs of creating that tent-pole product are just too high. This lesson is so easily seen through bad or just plain uninspired mobile design. Asking creative people to create uninspiring work is a fast track to mediocrity.

Here is a design solution: design for the best possible experience. Actually, don't just design for it: focus on creating the best possible experience with unwavering passion and commitment. Iterate, tweak, and fine-tune until you get it right. Anything less is simply unacceptable. Do not get hindered by the constraints of the technology. Phrases like "lowest common denominator" cannot be part of the designer's vocabulary.

Your design—no, your work of art—should serve as the shining example of what the experience should be, not what it can be. Trying to create a mobile design in the context of the device constraints isn't where you start; it is where you should end.

I think one of the greatest mistakes we in the mobile community make is being unwilling to or feeling incapable of thinking forward. The tendency to frame solutions in the past (past devices, past standards) applies only to those low-quality, something-for-everyone-but-getting-nothing tent-pole products. Great designs are not unlike great leaps forward in innovation. They come from shedding the baggage regarding how things are done and focus on giving people what they want or what they need.

The Elements of Mobile Design

As I wrote this chapter, I struggled with how to describe how to do design. I personally believe that good design requires three abilities: the first is a natural gift for being able to see visually how something should look that produces a desired emotion with the target audience. The second is the ability to manifest that vision into something for others to see, use, or participate in. The third is knowing how to utilize the medium to achieve your design goals.

Although I can't teach you how to remap your brain (at least not in the scope of this book) to be a designer, I can teach you how to at least think like a designer. Doing this involves knowing the six elements of mobile design that you need to consider, starting with the context and layering in visual elements or laying out content to achieve the design goal. Then, you need to understand how to use the specific tools to create mobile design, and finally, you need to understand the specific design considerations of the mobile medium.

Context

I discussed context in Chapter 4. I won't belabor the point except to say that context is core to the mobile experience. As the designer, it is your job to make sure that the user can figure out how to address context using your app. Make sure you do your homework to answer the following questions:

- Who are the users? What do you know about them? What type of behavior can you assume or predict about the users?

- What is happening? What are the circumstances in which the users will best absorb the content you intend to present?

- When will they interact? Are they at home and have large amounts of time? Are they at work where they have short periods of time? Will they have idle periods of time while waiting for a train, for example?

- Where are the users? Are they in a public space or a private space? Are they inside or outside? Is it day or is it night?

- Why will they use your app? What value will they gain from your content or services in their present situation?

- How are they using their mobile device? Is it held in their hand or in their pocket? How are they holding it? Open or closed? Portrait or landscape?

The answers to these questions will greatly affect the course of your design. Treat these questions as a checklist to your design from start to finish. They can provide not only great inspiration for design challenges, but justification for your design decisions later.

Message

Another design element is your message, or what you are trying to say about your site or application visually. One might also call it the "branding," although I see branding and messaging as two different things. Your message is the overall mental impression you create explicitly through visual design. I like to think of it as the holistic or at times instinctual reaction someone will have to your design. If you take a step back, and look at a design from a distance, what is your impression? Or conversely, look at a design for 30 seconds, and then put it down. What words would you use to describe the experience?

Branding shouldn't be confused with messaging. Branding is the impression your company name and logo gives—essentially, your reputation. Branding serves to reinforce the message with authority, not deliver it. In mobile, the opportunities for branding are limited, but the need for messaging is great. With such limited real estate, the users don't care about your brand, but they will care about the messaging, asking themselves questions like, "What can this do for me?" or "Why is this important to me?"

Your approach to the design will define that message and create expectations. A sparse, minimalist design with lots of whitespace will tell the user to expect a focus on content. A "heavy" design with use of dark colors and lots of graphics will tell the user to expect something more immersive.

For example, hold the book away from you and look at each of the designs in Figure 8-4; try not to focus too heavily on the content. What do each of these designs "say" to you?

Figure 8-4. What is the message for each of these designs?

Which of the following designs provide a message? What do they say to you?

Yahoo!

Yahoo! sort of delivers a message. This app provides a clean interface, putting a focus on search and location, using color to separate it from the news content. But

I'm not exactly sure what it is saying. Words you might use to describe the message are crisp, clean, and sharp.

ESPN

The ESPN site clearly is missing a message. It is heavily text-based, trying to put a lot of content above the fold, but doesn't exactly deliver a message of any kind. If you took out the ESPN logo, you likely would have indifferent expectations of this site; it could be about anything, as the design doesn't help set expectations for the user in any way. Words you might use to describe the message: bold, cluttered, and content-heavy.

Disney

Disney creates a message with its design. It gives you a lot to look at—probably too much—but it clearly tries to say that the company is about characters for a younger audience. Words you might use to describe the message: bold, busy, and disorienting.

Wikipedia

The Wikipedia design clearly establishes a message. With a prominent search and text-heavy layout featuring an article, you know what you are getting with this design. Words you might use to describe the message: clean, minimal, and text-heavy.

Amazon

Amazon sort of creates a message. Although there are some wasted opportunities above the fold with the odd ad placement, you can see that it is mostly about products (which is improved even more if you scroll down). Words you might use to describe the message: minimal but messy, product-heavy, and disorienting.

The words you might use to describe these designs might be completely different than mine—thus the beauty and the curse of visual design. The important thing isn't my opinion—it is the opinion of your user. Does the design convey the right message to your user in the right context? If you aren't sure, it might be a good time to find out.

Look and Feel

The concept of "look and feel" is an odd one, being subjective and hard to define. Typically, look and feel is used to describe appearance, as in "I want a clean look and feel" or "I want a usable look and feel." The problem is: as a mobile designer, what does it mean? And how is that different than messaging?

I think of look and feel in a literal sense, as something real and tactile that the users can "look" at, then "feel"—something they can touch or interact with. Look and feel is used to evoke action—how the user will use an interface. Messaging is holistic, as the expectation the users will have about how you will address their context. It is easy to confuse the two, because "feel" can be interpreted to mean our emotional reaction to design and the role of messaging.

I prefer to keep the concept of look and feel grounded in a tangible design, something I can clearly describe and show to users. I often find myself explaining the look and feel with the word "because," with a cause-and-effect rationale for design decisions, as in "The user will press this button because..." or "The user will go to this screen because..." followed by a reason why a button or control is designed a certain way.

Establishing a look and feel usually comes from wherever design inspiration comes from. However, your personal inspiration can be a hard thing to justify. Therefore we have "design patterns," or documented solutions to design problems, sometimes referred to as style guides. On large mobile projects or in companies with multiple designers, a style guide or pattern library is crucial, maintaining consistency in the look and feel and reducing the need for each design decision to be justified. For example, in Figure 8-5 you can see the site Pattern Tap (*http://patterntap.com*), which is a visual collection of many user interface patterns meant for websites and web applications, but there is no reason why it can't serve as inspiration for your mobile projects as well.

In Figure 8-6 you can see an example of a mobile design pattern at the Design4Mobile (*http://design4mobile.com*) pattern library.

Although a lot of elements go into making Apple's App Store successful, the most important design element is how it looks and feels. Apple includes a robust user interface tool that enables developers to use prebuilt components, supported with detailed Human Interface Guidelines (or HIG) of how to use them, similar to a pattern library. This means that a developer can just sit down and create an iPhone application that looks like it came from Apple in a matter of minutes. During the App Store submission process, Apple then ensures that the developer uses these tools correctly according to the HIG.

The look and feel can either be consistent with the stock user interface elements that Apple provides; they can be customized, often retaining the "spirit" of Apple's original design; or an entirely new look and feel can be defined—this approach is often used for immersive experiences.

The stock user experience that Apple provides is a great example of how look and feel works to supporting messaging. For the end user, the design sends a clear message: by using the same visual interface metaphors that Apple uses throughout the iPhone, I can expect the action, or how this control will behave, but I can also expect the same level of quality. This invokes the message of trust and quality in the application and in the platform as a whole. Apple isn't the first to use this shared look and feel model in mobile—in fact, it is incredibly common with most smartphone platforms—but they are surely making it incredibly successful, with a massive catalog of apps and the sales to support it.

My advice to would-be mobile designers is be creative and remember the context. Like in the early days of the Web, people tend to be skeptical about mobile experiences. The modal context of the user—in this case, what device he is using—should be considered during the design, as it will help to establish the user's expectations of the experience.

Figure 8-5. Pattern Tap shows a number of user interface patterns that help to establish look and feel

You can leverage this trust to your advantage, or you can strike out on your own and forge your own metaphors. As long as you know your users and the preferred mode of context, you can create a look and feel that is right for them.

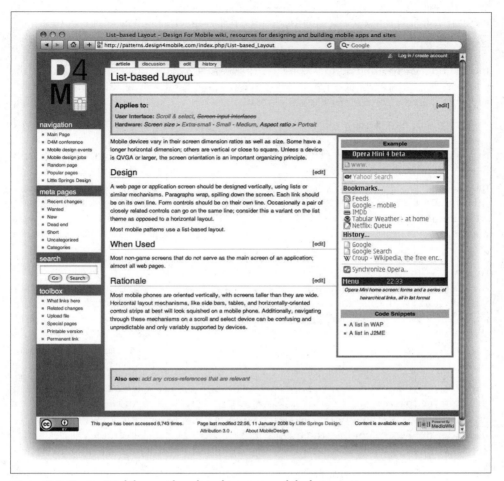

Figure 8-6. Design4Mobile provides a list of common mobile design patterns

Layout

Layout is an important design element, because it is how the user will visually process the page, but the structural and visual components of layout often get merged together, creating confusion and making your design more difficult to produce.

The first time layout should rear its head is during information architecture. In fact, I prefer to make about 90 percent of my layout decisions during the information architecture period. I ask myself questions like: where should the navigation go on the page or screen? What kind of navigation type should I use? Should I use tabs or a list? What about a sidebar for larger screens? All of these should be answered when defining the information architecture and before you begin to design.

Why define the layout before the mobile design? Design is just too subjective of an issue. If you are creating a design for anyone but yourself, chances are good that there will be multiple loosely-based-on-experience opinions that will be offered and debated. There is no right answer—only opinions and gut instincts. Plus, in corporate environments you have internal politics you have to consider, where the design opinions of the CEO or Chief Marketing Officer (CMO) might influence a design direction more than, say, the Creative Director or Design Director.

By defining design elements like layout prior to actually applying the look and feel, you can separate the discussion. As a self-taught designer, I started out in this business making designs for my own projects. I could just put pen to paper and tweak it to my heart's content. If I wanted to radically change the layout, I could. When I started my mobile design career with my first mobile company more than a decade ago, I realized that this approach didn't work. The majority of comments that reviewers would make were about the layout. They focused on the headers, the navigation, the footer, or how content blocks are laid out, and so on. But their feedback got muddied with the "look and feel, the colors, and other design elements."

Reviewers do make remarks like "I like the navigation list, but can you make it look more raised?" Most designers don't hear that; they hear "The navigation isn't right, do it again." But, with this kind of feedback, there are two important pieces of information about different types of design. First, there is confirmation that the navigation and layout are correct. Second, there is a question about the "look and feel." Because designers hear "Do it again," they typically redo the layout, even though it was actually fine.

Creating mobile designs in an environment with multiple reviewers is all about getting the right feedback at the right time. Your job is to create a manifestation of a shared vision. Layout is one of the elements you can present early on and discuss independently. People confuse the quality and fidelity of your deliverables as design. By keeping it basic, you don't risk having reviewers confuse professionalism with design.

The irony is that as I become more adept at defining layouts, I make them of increasingly lower fidelity. For example, when I show my mobile design layouts as wireframes during the information architecture phase, I intentionally present them on blueprint paper, using handwriting fonts for my annotations (Figure 8-7). It also helps to say that this is not a design, it is a layout, so please give me feedback on the layout.

Different layouts for different devices

The second part of layout design is how to visually represent content. In mobile design, the primary content element you deal with the is navigation. Whether you are designing a site or app, you need to provide users with methods of performing tasks, navigating to other pages, or reading and interacting with content. This can vary, depending on the devices you support.

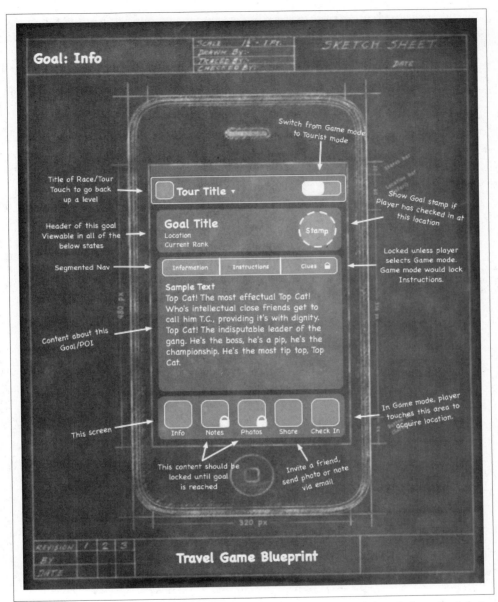

Figure 8-7. Using a low-fidelity wireframe to define the layout design element before visual design begins

There are two distinct types of navigation layouts for mobile devices: touch and scroll. With touch, you literally point to where you want to go; therefore, navigation can be anywhere on the screen. But we tend to see most of the primary actions or navigation areas living at the bottom of the screen and secondary actions living at the top of the screen, with the area in between serving as the content area, like what is shown in Figure 8-8.

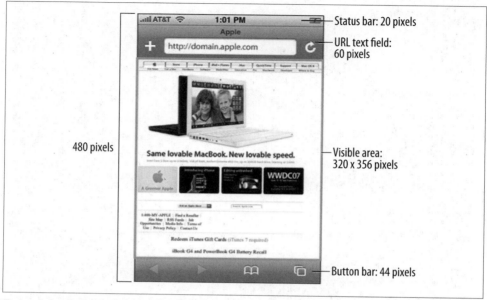

Figure 8-8. iPhone HIG, showing the layout dimensions of Safari on the iPhone

This is the opposite of the scroll navigation type, where the device's D-pad is used to go left, right, up, or down. When designing for this type of device, the primary and often the secondary actions should live at the top of the screen. This is so the user doesn't have to press down dozens of times to get to the important stuff. In Figure 8-9, you can actually see by the bold outline that the first item selected on the screen is the link around the logo.

When dealing with scroll navigation, you also have to make the choice of whether to display navigation horizontally or vertically. Visually, horizontally makes a bit more sense, but when you consider that it forces the user to awkwardly move left and right, it can quickly become a bit cumbersome for the user to deal with. There is no right or wrong way to do it, but my advice is just to try and keep it as simple as possible.

Fixed versus fluid

Another layout consideration is how your design will scale as the device orientation changes, for example if the device is rotated from portrait mode to landscape and vice versa. This is typically described as either being fixed (a set number of pixels wide), or fluid (having the ability to scale to the full width of the screen regardless of the device orientation).

Figure 8-9. Example layout of a scroll-based application, where the user had to press the D-pad past each link to scroll the page

Orientation switching has become commonplace in mobile devices, and your design should always provide the user with a means to scale the interface to take full advantage of screen real estate.

Color

The fifth design element, color, is hard to talk about in a black-and-white book. Maybe it is fitting, because it wasn't that long ago that mobile screens were available only in black and white (well, technically, it was black on a green screen). These days, we have nearly the entire spectrum of colors to choose from for mobile designs.

The most common obstacle you encounter when dealing with color is mobile screens, which come in a number of different color or bit depths, meaning the number of bits (binary digits) used to represent the color of a single pixel in a bitmapped image. When complex designs are displayed on different mobile devices, the limited color depth on one device can cause banding, or unwanted posterization in the image.

For an example of posterization, the technical term for when the gradation of tone is replaced with regions of fewer tones, see in Figure 8-10 how dramatically the color depth can affect the quality of a photo or gradient, producing banding in several parts in the image.

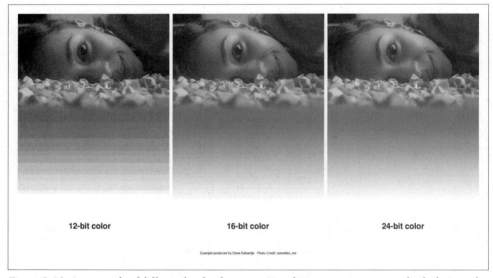

Figure 8-10. An example of different levels of posterization that can occur across multiple device color depths

Different devices have different color depths. In Table 8-1, you can see the supported colors and a few example devices.

Table 8-1. Supported colors and example devices

Bit depth	Supported colors	Description	Example devices
12-bit	4,096 colors	Used with older phones; dithering artifacts in photos can easily be seen.	Nokia 6800
16-bit	65,536 colors	Also known as HighColor; very common in today's mobile devices. Can cause some banding and dithering artifacts in some designs.	HTC G1, BlackBerry Bold 9000, Nokia 6620

Bit depth	Supported colors	Description	Example devices
18-bit	262,144 colors	Used in mobile devices to offer Truecolor (see following entry) levels through dithering. Limited banding may be seen.	Samsung Alias, Sony Ericsson TM506
24-bit	16.7 million colors	Also known as Truecolor; supports millions of colors and produces little banding.	iPhone, Palm Prē, Nokia N97

The psychology of color

People respond to different colors differently. It is fairly well known that different colors produce different emotions in people, but surprisingly few talk about it outside of art school. Thinking about the emotions that colors evoke in people is an important aspect of mobile design, which is such a personal medium that tends to be used in personal ways. Using the right colors can be useful for delivering the right message and setting expectations.

One of the examples I used earlier was the ESPN mobile site, which uses a bold red header to create a stark and prominent tone to the design. But what does that say about ESPN? What does it tell the user about the experience?

For the purposes of reference, Table 8-2 provides some of the characteristics of various colors that naturally evoke certain emotions in people.

Table 8-2. Color characteristics

Color	Represents
White	Light, reverence, purity, truth, snow, peace, innocence, cleanliness, simplicity, security, humility, sterility, winter, coldness, surrender, fearfulness, lack of imagination, air, death (in Eastern cultures), life, marriage (in Western cultures), hope, bland
Black	Absence, modernity, power, sophistication, formality, elegance, wealth, mystery, style, evil, death (in Western cultures), fear, seriousness, conventionality, rebellion, anarchism, unity, sorrow, professionalism
Gray	Elegance, humility, respect, reverence, stability, subtlety, wisdom, old age, pessimism, boredom, decay, decrepitude, dullness, pollution, urban sprawl, strong emotions, balance, neutrality, mourning, formality
Yellow	Sunlight, joy, happiness, earth, optimism, intelligence, idealism, wealth (gold), summer, hope, air, liberalism, cowardice, illness (quarantine), fear, hazards, dishonesty, avarice, weakness, greed, decay or aging, femininity, gladness, sociability, friendship
Green	Intelligence, nature, spring, fertility, youth, environment, wealth, money (U.S.), good luck, vigor, generosity, go, grass, aggression, coldness, jealousy, disgrace (China), illness, greed, drug culture, corruption (North Africa), life eternal, air, earth (classical element), sincerity, renewal, natural abundance, growth
Blue	Seas, men, productiveness, interiors, skies, peace, unity, harmony, tranquility, calmness, trust, coolness, confidence, conservatism, water, ice, loyalty, dependability, cleanliness, technology, winter, depression, coldness, idealism, air, wisdom, royalty, nobility, Earth (planet), strength, steadfastness, light, friendliness, peace, truthfulness, love, liberalism (U.S. politics), and conservatism (UK, Canadian, and European politics)
Violet	Nobility, envy, sensuality, spirituality, creativity, wealth, royalty, ceremony, mystery, wisdom, enlightenment, arrogance, flamboyance, gaudiness, mourning, exaggeration, profanity, bisexuality, confusion, pride

Color	Represents
Red	Passion, strength, energy, fire, sex, love, romance, excitement, speed, heat, arrogance, ambition, leadership, masculinity, power, danger, gaudiness, blood, war, anger, revolution, radicalism, aggression, respect, martyrs, conservatism (U.S. politics), Liberalism (Canadian politics), wealth (China), and marriage (India)
Orange	Energy, enthusiasm, balance, happiness, heat, fire, flamboyance, playfulness, aggression, arrogance, gaudiness, over-emotion, warning, danger, autumn, desire
Pink	Spring, gratitude, appreciation, admiration, sympathy, socialism, femininity, health, love, romance, marriage, joy, flirtatiousness, innocence and child-like qualities
Brown	Calm, boldness, depth, nature, richness, rustic things, stability, tradition, anachronism, boorishness, dirt, dullness, heaviness, poverty, roughness, earth

Note what some of the different colors can mean in different cultures. In some cases, the color you use can have opposing meanings in different cultures. This is something to consider when thinking of deploying your mobile experience to countries with the highest number of mobile devices, such as China or India.

Color palettes

Defining color palettes can be useful for maintaining a consistent use of color in your mobile design. Color palettes typically consist of a predefined number of colors to use throughout the design. Selecting what colors to use varies from designer to designer, each having different techniques and strategies for deciding on the colors. I've found that I use three basic ways to define a color palette:

Sequential

In this case, there are primary, secondary, and tertiary colors. Often the primary color is reserved as the "brand" color or the color that most closely resembles the brand's meaning. The secondary and tertiary colors are often complementary colors that I select using a color wheel.

Adaptive

An adaptive palette is one in which you leverage the most common colors present in a supporting graphic or image. When creating a design that is meant to look native on the device, I use an adaptive palette to make sure that my colors are consistent with the target mobile platform.

Inspired

This is a design that is created from the great pieces of design you might see online, as shown in Figure 8-11, or offline, in which a picture of the design might inspire you. This could be anything from an old poster in an alley, a business card, or some packaging. When I sit down with a new design, I thumb through some of materials to create an inspired palette. Like with the adaptive palette, you actually extract the colors from the source image, though you should never ever use the source material in a design.

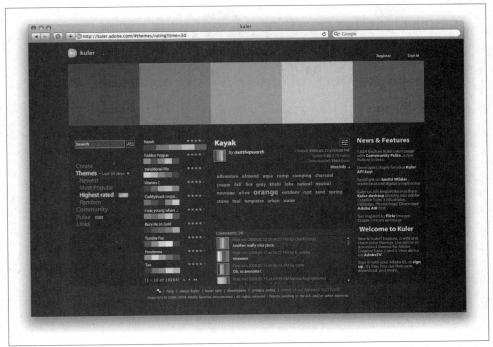

Figure 8-11. Adobe Kuler (http://kuler.adobe.com), a site that enables designers to share and use different color palettes

Typography

The sixth element of mobile design is typography, which in the past would bring to mind the famous statement by Henry Ford:

> Any customer can have a car painted any color that he wants so long as it is black.

Traditionally in mobile design, you had only one typeface that you could use (Figure 8-12), and that was the device font. The only control over the presentation was the size.

As devices improved, so did their fonts. Higher-resolution screens allowed for a more robust catalog of fonts than just the device font. First, let's understand how mobile screens work.

Subpixels and pixel density

There seem to be two basic approaches to how type is rendered on mobile screens: using subpixel-based screens or having a greater pixel density or pixels per inch (PPI).

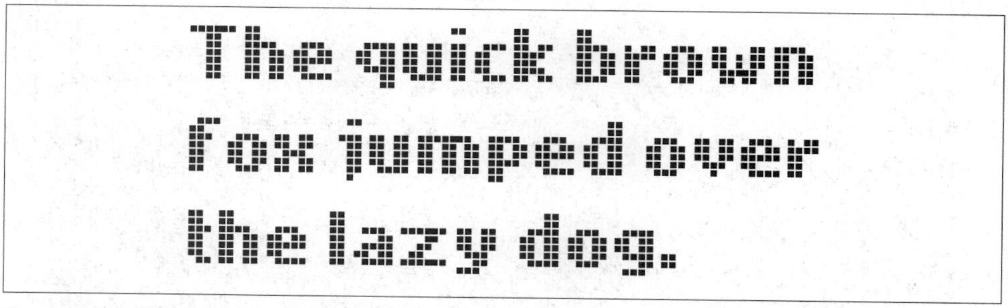

Figure 8-12. What most mobile designers think of when it comes to mobile typography

A subpixel is the division of each pixel into a red, green, and blue (or RGB) unit at a microscopic level, enabling a greater level of antialiasing for each font character or glyph. The addition of these RGB subpixels enables the eye to see greater variations of gray, creating sharper antialiasing and crisp text.

In Figure 8-13, you can see three examples of text rendering. The first line shows a simple black and white example, the second shows text with grayscale antialiasing, and the third line shows how text on a subpixel display would render.

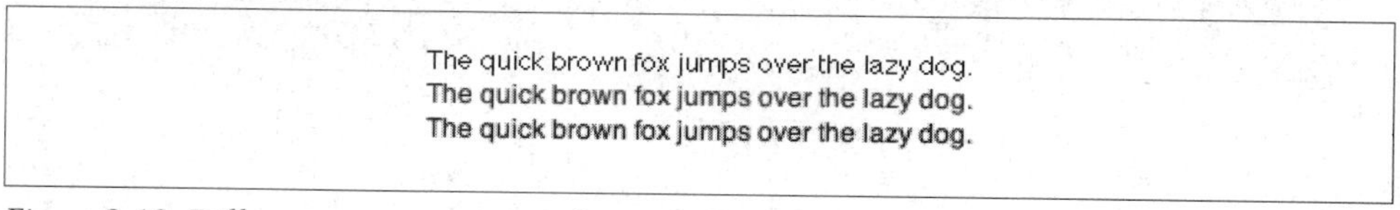

The quick brown fox jumps over the lazy dog.
The quick brown fox jumps over the lazy dog.
The quick brown fox jumps over the lazy dog.

Figure 8-13. Different ways text can render on mobile screens

The Microsoft Windows Mobile platform uses the subpixel technique with its Clear-Type technology, as shown in Figure 8-14.

The second approach is to use a great pixel density, or pixels per inch. We often refer to screens by either their actual physical dimensions ("I have a 15.4-inch laptop screen") or their pixel dimensions, or resolution ("The resolution of my laptop is 1440×900 pixels"). The pixel density is determined by dividing the width of the display area in pixels by the width of the display area in inches. So the pixel density for my 15.4-inch laptop would be 110 PPI. In comparison, a 1080p HD television has a PPI of 52.

As this applies to mobile devices, the higher the density of pixels, the sharper the screen appears to the naked eye. This guideline especially applies to type, meaning that as text is antialiased on a screen with a high density of tiny pixels, the glyph appears sharper to the eye. Some mobile screens have both a high PPI and subpixel technology, though these are unnecessary together.

Figure 8-14. Microsoft ClearType using subpixels to display sharp text

Table 8-3 provides the dimensions and PPI for a few mobile devices.

Table 8-3. Dimensions and PPI for some mobile devices

Mobile device	Diagonal	Pixels	PPI
Nokia N95	2.6"	240×320	153
Apple iPhone 3G	3.5"	320×480	163
Amazon Kindle	6.0"	600×800	167
HTC Dream	3.2"	320×480	181
Sony Ericsson W880i	1.8"	240×320	222
Nokia N80	2.1"	352×416	256

Type options

Fortunately, today's mobile devices have a few more options than a single typeface, but the options are still fairly limited. Coming from web design, where we have a dozen or so type options, the limited choices available in mobile design won't come as a big surprise. Essentially, you have a few variations of serif, sans-serif, and monospace fonts, and depending on the platform, maybe a few custom fonts (Figure 8-15).

Figure 8-15. Options in typography increase as the devices become more sophisticated

In researching this book, I scoured the Web and tapped my mobile community resources to find a list of the typefaces that are included in each of the major device platforms, but I could only come up with a few—nothing close to a complete list. This goes to show how far behind mobile typography is, that designers don't even have a basic list to work from.

Therefore, when creating mobile designs for either web or native experiences, my advice is to stick with either the default device font, or web-safe fonts—your basic serif variants like Times New Roman and Georgia or sans-serif typefaces like Helvetica, Arial, or Verdana.

Font replacement

The ability to use typefaces that are not already loaded on the device varies from model to model and your chosen platform. Some device APIs will allow you to load a typeface into your native application. Some mobile web browsers support various forms of font replacement; the two most common are sIFR and Cufon. sIFR uses Flash to replace HTML text with a Flash representation of the text, but the device of course has to support Flash. Cufon uses JavaScript and the canvas element draws the glyphs in the browser, but the device of course needs to support both JavaScript and the canvas element.

In addition, the `@font-face` CSS rule allows for a typeface file to be referenced and loaded into the browser, but a license for web use is usually not granted by type foundries.

Readability

The most important role of typography in mobile design is to provide the user with excellent readability, or the ability to clearly follow lines of text with the eye and not lose one's place or become disoriented, as shown in Figure 8-16. This can be done by following these six simple rules:

Use a high-contrast typeface
> Remember that mobile devices are usually used outside. Having a high-contrast typeface with regard to the background will increase visibility and readability.

Use the right typeface
> The type of typeface you use tells the user what to expect. For example, a sans-serif font is common in navigation or compact areas, whereas serif typefaces come in handy for lengthy or dense content areas.

Provide decent leading (rhymes with "heading") or line spacing
> Mobile screens are often held 10–12" away from the eye, which can make tracking each line difficult. Increase the leading to avoid having the users lose their place.

Leave space on the right and left of each line; don't crowd the screen
> Most mobile frameworks give you full access to the screen, meaning that you normally need to provide some spacing between the right and left side of the screen's edge and your text—not much, typically about three to four character widths.

Generously utilize headings
> Break the content up in the screen, using text-based headings to indicate to the user what is to come. Using different typefaces, color, and emphasis in headings can also help create a readable page.

Use short paragraphs
> Like on the Web, keep paragraphs short, using no more than two to three sentences per paragraph.

Figure 8-16. Classics, an iPhone application designed with readability and typography in mind

Graphics

The final design element is graphics, or the images that are used to establish or aid a visual experience. Graphics can be used to supplement the look and feel, or as content displayed inline with the text.

For example, in Figure 8-17, you can see Ribot's Little Spender application for the iPhone and the S60 platform. The use of graphical icons in the iPhone experience helps to establish a visual language for the user to interact with to quickly categorize entries. On the S60 application, the wallet photo in the upper-right corner helps communicate the message of the application to the user.

Iconography

The most common form of graphics used in mobile design is icons. Iconography is useful to communicate ideas and actions to users in a constrained visual space. The challenge is making sure that the meaning of the icon is clear to the user. For example, looking at Figure 8-18, you can see some helpful icons that clearly communicate an idea and some perplexing icons that leave you scratching your head.

Figure 8-17. Ribot's Little Spender application uses graphics to define the experience

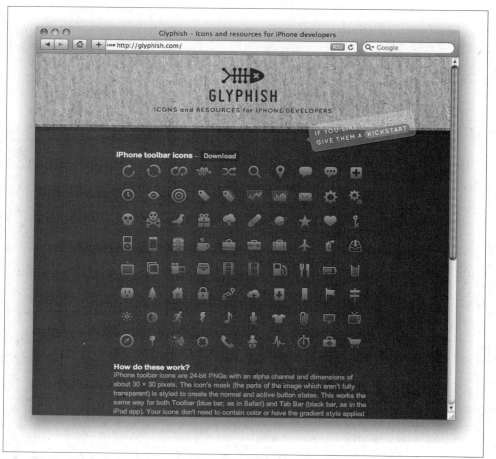

Figure 8-18. Glyphish (http://glyphish.com) provides free iPhone icons

Photos and images

Photos and images are used to add meaning to content, often by showing a visual display of a concept, or to add meaning to a design. Using photos and images isn't as common in mobile design as you might think. Because images have a defined height and width, they need to be scaled to the appropriate device size, either by the server, using a content adaptation model, or using the resizing properties of the device. In the latter approach, this can have a cost in performance. Loading larger images takes longer and therefore costs the user more.

Using graphics to add meaning to a design can be a useful visual, but you can encounter issues regarding how that image will display in a flexible UI—for example, when the device orientation is changed. In Figure 8-19, you can see how the pig graphic is designed to be positioned to the right regardless of the device orientation.

Figure 8-19. Using graphics in multiple device orientations

Mobile Design Tools

As I mentioned earlier, mobile design requires understanding the design elements and specific tools. The closest thing to a common design tool is Adobe Photoshop, though each framework has a different method of implementing the design into the application. Some frameworks provide a complete interface toolkit, allowing designers or developers to simply piece together the interface, while others leave it to the designer to define from scratch.

In Table 8-4, you can see each of the design tools and what interface toolkits are available for it.

Table 8-4. Design tools and interface toolkits

Mobile framework	Design tool	Interface toolkits
Java ME	Photoshop, NetBeans	JavaFX, Capuchin
BREW	Photoshop, Flash	BREW UI Toolkit, uiOne, Flash
Flash Lite	Flash	Flash Lite
iPhone	Photoshop, Interface Builder	iPhone SDK

Mobile framework	Design tool	Interface toolkits
Android	Photoshop, XML-based themes	Android SDK
Palm webOS	Photoshop, HTML, CSS, and JavaScript	Mojo SDK
Mobile web	Photoshop, HTML, CSS, and JavaScript	W3C Mobile Web Best Practices
Mobile widgets	Photoshop, HTML, CSS, and JavaScript	Opera Widget SDK, Nokia Web Runtime
Mobile web apps	Photoshop, HTML, CSS, and JavaScript	iUI, jQTouch, W3C Mobile Web App Best Practices

Designing for the Right Device

With the best possible experience at hand, take a moment to relish it. Remind yourself that you are working with a rapidly evolving medium and though it might not be possible for every user to experience things exactly the way you've intended, you've set the tone and the vision for how the application should look. The truly skilled designer doesn't create just one product—she translates ideas into experiences. The spirit of your design should be able to be adapted to multiple devices.

Now is the time to ask, "What device suits this design best? What market niche would appreciate it most? What devices are the most popular within that niche?" The days of tent-poles are gone. Focus instead on getting your best possible experience to the market that will appreciate it most. It might not be the largest or best long-term market, but what you will learn from the best possible scenario will tell you volumes about your mobile product's potential for success or failure. You will learn which devices you need to design for, what users really want, and how well your design works in the mobile context.

This knowledge will help you develop your porting and/or adaptation strategy, the most expensive and riskiest part of the mobile equation. For example, if you know that 30 percent of your users have iPhones, then that is a market you can exploit to your advantage. iPhone users consume more mobile content and products than the average mobile user. This platform has an easy-to-learn framework and excellent documentation, for both web and native products, and an excellent display and performance means. Although iPhone users might not be the majority of your market, the ability to create the best possible design and get it in front of those users presents the least expensive product to produce with the lowest risk.

With a successful single device launch, you can start to adapt designs from the best possible experience to the second best possible experience, then the third, and fourth, and so on. The best possible experience is how it should be, so it serves as a reference point for how we will adapt the experience to suit more devices.

Designing for Different Screen Sizes

Mobile devices come in all shapes and sizes. Choice is great for consumers, but bad for design. It can be incredibly difficult to create that best possible experience for a plethora of different screen sizes. For example, your typical feature phone might only be 140 pixels wide, whereas your higher-end smartphone might be three to four times wider.

Landscape or portrait? Fixed width or fluid? Do you use one column or two? These are common questions that come up when thinking about your design on multiple screen sizes. The bad news is that there is no simple answer. How you design each screen of content depends on the scope of devices you look to support, your content, and what type of experience you are looking to provide. The good news is that the vast majority of mobile device screens share the same vertical or portrait orientation, even though they vary greatly in dimension, as shown in Figure 8-20.

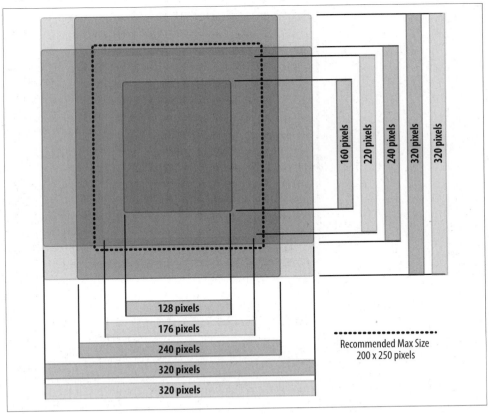

Figure 8-20. Comparing the various screen sizes

Of course, there are some devices by default in a horizontal orientation, and many smartphones that can switch between the two orientations, but most people use their mobile devices in portrait mode. This is a big shift in thinking if you are coming from interactive design, as up to this point, screens have been getting wider, not taller.

For years now, we've become used to placing less-crucial information along the sides of web pages. In software, tasks flow from left to right. With vertical designs, the goal is to think of your design as a cascade of content from top to bottom (Figure 8-21), similar to a newspaper. The most contextual information lives at the top, and the content consumes the majority of the screen. Any exit points live at the bottom. Mobile is no different.

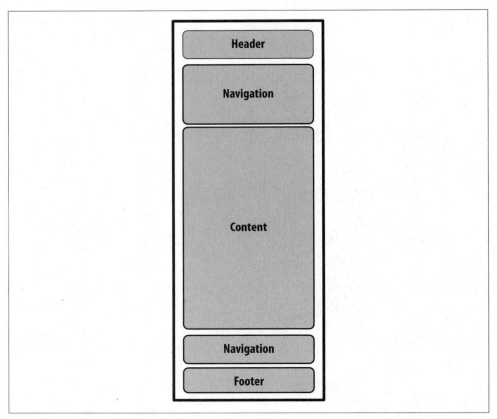

Figure 8-21. The typical flow of information on mobile devices

The greatest challenge to creating a design that works well on multiple screen sizes is filling the width. For content-heavy sites and applications, the width of mobile devices is almost the perfect readability, presenting not too many words per line of text. The problem is when you have to present a number of tasks or actions. The easiest and most compatible way is to present a stacked list of links or buttons, basically one action per line. It isn't the most effective use of space, but presenting too many actions on the horizontal axis quickly clutters the design—not to mention that it is more difficult to adapt to other devices.

Unfortunately, it isn't always reasonable to implement fluid or flexible designs that stretch to fit the width of the screen. Although most mobile web browsers and device framework APIs enable it in principle, its execution across multiple devices is a little anticlimatic. Mobile websites usually employ a fixed-width layout for the lowest common denominator, and native applications are often resized for multiple screen sizes during development.

As devices get larger, denser screens, you will see an increase in the use of touch, forcing the size of content to increase to fingertip size—typically 40 pixels wide and 40 pixels tall (Figure 8-22). This actually solves part of the horizontal axis problem, simply by making content larger for larger screens. Ironically, you can fit almost the same amount of usable content in an iPhone as you can a lower-end device.

Figure 8-22. The iPhone calculator application uses common fingertip-size controls

Obviously, you can fit a lot more on screen with more advanced devices, but you want to avoid forcing the user to zoom in and out of your interfaces.

## CHAPTER 9

Mobile Web Apps Versus Native Applications

When should you make a mobile web application versus a native mobile application, or a downloadable application designed for a specific platform, like a Java application or an iPhone? It wasn't that long ago that people didn't even bother asking the question; every mobile application was native.

The mobile web historically has been so horrendous to deal with that the only way to create a compelling experience was to go native. Assuming that native applications will immediately create compelling experience is misleading, though, as dealing with native applications across multiple platforms isn't exactly a walk in the park, either. Device fragmentation exists across devices, because if you are dealing with native APIs or mobile web browsers, it is an obstacle that cannot be avoided.

But like most things in the mobile ecosystem, the question of which path to choose comes down to money. Native applications can produce a lot of short-term revenue for developers and operators alike, especially with the proliferation of app stores. Though traffic to mobile websites and web applications produce a sizable chunk of operator revenue in data charges, developers assume they get nothing. This isn't to say that making money from the mobile web can't be done; it's just that few know how to choose the right path to monetization.

In this chapter, I discuss the reasons why you should create a native application, but first, let's start by setting the stage with some historical context.

The Ubiquity Principle

Let me start by clearly stating where I stand on this issue: I believe that the mobile web is the only long-term commercially viable content platform for mobile devices. I have four key reasons to support this belief: fragmentation, the Web, control, and consumer expectations.

Fragmentation

First of all, we already know that mobile is a much larger playing field than desktop computing, but there is currently no economically feasible means to create native applications that can support the majority of the market. There isn't just a handful of platforms to contend with or a clear market leader, but literally hundreds (when you include all the variations), with no one vendor able to firmly claim itself king. Getting your application on one platform is a snap, but getting it on two is a challenge, five a costly headache, and supporting fifty virtually impossible.

The Web

> Anyone who's betting against the Web right now is an idiot.
>
> —Daniel Appelquist, Co-Chair W3C Mobile Web Initiative

The overall technology market is going to the Web. It is a highly vetted consumer medium that offers many pros and few cons. It is the only medium for information, applications, and services that has gone the distance for the last 15 years. With a new focus on advanced desktop web browser technology, which is poised to become more than just rendering text, the only native application that matters is the browser. The majority of digital innovation is occurring using the technologies of the Web, making the browser the central delivery mechanism for the applications and services of tomorrow. As more mobile browsers add services to detect location, acceleration, or use of the hardware, I predict that the need for native applications will be reduced to specific uses that really need the full capacity of the device, like games.

Control

Mobile application distribution cannot and will likely never be under the control of the developer. In other words, mobile application vendors always have to rely on middlemen to get their products to market and take a slice of their profits. This has been the case since the beginning of downloadable mobile applications, when they were under the tight control of the operator.

These days, we see that control shifting to the device and platform makers, making getting applications to market easier. But do not be fooled—the model is exactly the same. The purpose of your product is only to service them, boosting their bottom line. The lack of control or even influence of your primary distribution channel puts your product at risk almost from the start. Your product can be shelved after just a few days in the spotlight.

Without control, this reason alone means that the funding of creating mobile applications will always remain a small, high-risk investment. Without the funding to help weather market fluctuations or the willingness to invest in the truly innovative products, developers will be forced to continue knocking out small, pointless applications aimed at short-term revenue gain—novelty products that wear off quickly.

Consumer Expectations

My fourth belief is that consumers expect things to just work, and rightfully so. The challenge with native mobile applications is that the consumer may see an application that might look appealing to him, but if it isn't supported for his particular device, then not only is the sale lost, but often the customer is lost for good. This is one of the reasons that operators usually require applications sold on their marketplace to support their top 10 to 15 devices.

From the consumer's perspective, he spends good money on a device and wants content to support it. The lack of available content lowers the perceived value of the device. Consumers don't care what device or platform they have; they just want to participate in the same content and services that their friends are using. Because cross-platform support is so challenging, that is hardly ever the possible.

You can see these traffic trends in just about all mobile marketplaces. Numerous visits occur when a consumer has purchased a brand-new device, but then visits drop off precipitately in just days. After the device is just a month old, the likelihood that the user will return for more content is slim to none.

Ubiquity in the Mobile Web

The mobile web is the only platform that is available and works across all mobile devices, sharing the same set of standards and protocols with each other as well as the desktop web. The mobile web is also the only mobile distribution channel available to developers that they can control. It is the best way to bridge short, context-based mobile interactions with longer, desktop-based tasks.

The mobile web is the easiest platform to learn, the cheapest to produce, the most standardized, the most available, and the easiest to distribute. I call this the Ubiquity Principle: easier-to-produce quality content and services for the largest available market will always win.

The key word is "quality," which the mobile web hasn't had a lot of over the years. Although the mobile web has its own challenges with device fragmentation, the level of complexity to adapt to these challenges to produce the best possible experience is far lower than with native applications. In addition, these challenges are going away quickly and will be inconsequential in just a few years' time.

When to Make a Native Application

I'm obviously a big supporter of the mobile web; however, I am the first one to admit that making a native application can be the best thing for a product. This is usually true when you need to take advantage of the features of a device that a mobile web browser does not allow.

The following sections discuss some of the key features you may be considering that almost immediately point toward creating a native application.

Charging for It

Nowhere is it written that you can't charge people to use a mobile web app, but for some reason, people think you can't or shouldn't. Historically, charging for things on mobile devices has come down to two obstacles:

- Entering a credit card number in the mobile device is cumbersome and can be insecure on older devices. Typically, if you want to charge for it, you set up an arrangement with operators to do Billing on Behalf of (BoBo) so purchases just show up on the subscriber's bill. This means you need to have BoBo arrangements with every carrier your application is on, which is its own headache. This is usually the preferred method, because a lot of people with phones don't have credit cards, such as young people.

  Another route is storing subscribers' credit card information on a secure website. The user is then able to log into her mobile profile and make purchases against it. This two-step process is less than ideal and basically means that people can't just go and buy from their device.

- Operators want their cut. A native mobile application directory, either through the operator or through a device maker, always includes a means to collect payment for it. They typically just take their cut and pass on the rest of it to you, but it means that you must work within the rules and regulations of their marketplace. Getting onto the operator marketplace has historically been a colossal effort, requiring full-time staff just to negotiate terms. Since those days, device maker markets have made it much easier to get products to market, but are plagued by most, if not all, of the same problems and red tape.

  Applications and services that encroach on the operator or device maker's business model will likely be blocked or removed from sale. Mobile websites that earn too much money without the carrier getting a percentage have been shut down in the past, but you don't hear many of those stories these days.

The bottom line is that if you want to charge for a native application, you are going to have to work within the rules of the marketplace and give up a decent percentage of your profits.

Creating a Game

If your goal is to create a mobile game—one of the biggest mobile content markets—then you need to create a native application. Games are resource-intensive and almost always require the use of a device or platform API. Although there have been quite a few compelling games created using just web technologies, the competition in the native application game market is just too stiff. When users launch a game, they have some expectations of what it is going to look and act like. The mobile web is close to providing an analog experience, but is not quite there yet.

When making mobile games, you need to carefully consider which platforms to support. Luckily, there are a variety of game porting houses that can help get your game onto multiple platforms, but as you might imagine, it can be a costly endeavor in terms of money and time.

Using Specific Locations

The next feature is location, or being able to detect the users' locations by GPS or cell tower triangulation for the purpose of presenting users with information based on their current location. Traditionally, the only way to access users' locations is through native application APIs, but the W3C Geolocation API is quickly being incorporated into the most popular mobile browsers. Devices that run WebKit, like the iPhone or Android devices, as well as devices that run the Opera or Mozilla browsers, will all have the ability to detect user's locations.

I believe that location is the holy grail of the Web. If it can be made available to the web browser, then web publishers can begin to use location and context in new and interesting ways. Although it is not a technical hurdle, the issue has mostly been with privacy. We like to think of the web browser as an anonymous portal into the World Wide Web. Adding location means sharing sensitive information with websites, which could actually be quite dangerous. But for all location-aware applications, providing your location requires your direct authorization from the device for each instance, and you have the option to disable it entirely.

Using Cameras

The camera is another device function that can come in handy in your applications. Traditionally, mobile MMS (Multimedia Messaging Service) is used to handle mobile photo interactions. In other words, you take a photo, send it via MMS to a shortcode, then a server somewhere does something with the photo and sends an alert to you when it is done. This process is complicated, time-consuming, and fairly unreliable.

With access to the camera, native application developers can simplify the task to just taking a photo from within an application. The user then can do limited processing on the client side, sending to a server only if necessary, through a much more reliable HTTP connection. The W3C is working on an API to allow for camera access, but it has yet to be incorporated by the browser vendors.

The camera is useful in several types of mobile applications, from sharing snapshots or videos of friends to capturing important events as they happen and recording visual information, such as an important advertisement or sign found while out and about. The camera can even be used to process bar codes and then to redeem promotions. Not long from now, we could see cameras that are able to translate different printed languages just by holding our mobile camera up to a sign—a technology that is already popular in Japan.

Using Accelerometers

A popular feature in more recent mobile devices is the addition of an accelerometer, a small instrument that measures physical acceleration and gravity and sends data back to the device. The most common use of an accelerometer is to detect when the device is physically rotated, adjusting the display from vertical to horizontal orientation (or vice versa).

Using the accelerometer can be a benefit to mobile users, enabling them to interact with a device in a more natural way; being that the device is likely held in the hand, it can adjust content to suit its physical orientation, like rotating the screen, or detecting physical movement, and can therefore have limited prediction of the users' context. A simple example might be that if the user is walking, detected by slight movement or velocity, the user interface might display larger text than normal to make the experience easier for users on the move.

However, developers should be cautioned against relying too heavily on the accelerometer for frivolous interactions. Every mobile interaction should pass the "transit test." You should always assume that the user will interact with a mobile device the same way she would if she were sitting on a crowded bus or train. Ask yourself what the likelihood is that the user will shake the phone while standing in a packed subway or riding in a car. Typically, the likely answer is slim to none. So be sure to always include an alternative way to perform the same task that attracts less attention.

Accessing the Filesystems

Another reason you would want to create a native application is if you want to use the data stored on the device itself. This might be the user's address book, photos, an email message, or even data from another application.

Such filesystem access is obviously a big security and privacy issue. An errant application potentially could alter or delete your data, or worse. An infected application could use your contacts and constant connection to spread a virus to multiple phones, something that occurred quite often before widespread adoption of mobile application certification.

On the other hand, mobile devices are becoming highly personal, mobile computers that store an increasing amount of content and information about their owners and the owners' friends and business contacts. The idea of leveraging this information across applications is appealing. Though not without risk, using stored data is a powerful way to present contextually relevant information to the user.

Developers should be cautioned to use stored information only in limited doses. We've seen a trend with applications that leverage too much user data without the user's express permission as being mistakenly perceived as spamming or phishing information, even though the application was actually attempting to perform a valuable service. False perceptions of your application can and will significantly affect your ability to distribute it, either in loss of direct sales or possibly by being pulled altogether, if the operator receives enough complaints.

The rule of thumb here is to never do anything with the users' data without their express permission to do so: a precaution far too many mobile applications skip.

Again, we are seeing some progress by the W3C to establish a standard API for mobile vendors to follow, but we aren't there yet.

Offline Users

The final reason to make a native application is because you know the user is likely to be offline or out of range of a mobile network. For those of us who live in the city, that may seem like a rare occasion. Even for those who live in more rural areas, network dead spots are becoming increasingly rare. But going periods without a connection does of course happen frequently and your application should be designed to take this into account.

Think about the context of the user and when he is likely to use your application. The likelihood that a mobile game will be played on an airplane would probably be pretty high. A trail map application will likely be used in more remote areas, with less coverage. A mobile travel guide will probably be used in a foreign network, where roaming and international fees could occur. Each of these applications should have an offline mode in which the user can still perform the most common tasks of the application without the need of a wireless connection.

The mobile browsers that support HTML5 actually include the capability to create offline apps today, but it isn't made apparent to the user. As support for offline storage increases across multiple mobile browsers, we need to define metaphors and conventions for the users to know when their mobile web application will work when the device is offline.

I find that many native applications assume reliable network connectivity far too often. Applications are often designed under the best of circumstances, in a closed environment with a healthy wireless signal and a fast network. Mobile devices by nature move from the best of circumstances to the worst quickly. Native applications should always be tested under the worst possible conditions. For example, what happens if the user starts a task with a full signal, but tries to complete it with no signal?

Users don't think about being online or offline when they load native applications—they just expect them to work. It's our job to make sure that they do.

When to Make a Mobile Web Application

I believe that unless your application meets one of these native application criteria, you should not create a native application, but should instead focus on building a mobile web application. Like I said before, I'm a big fan of native applications and I feel that there are a lot of great innovative and market opportunities here, but mobile web apps are the only long-term viable platform for mobile content, services, and applications.

Native applications don't service the user better in any significant way; they only add cost to your project, decrease your distribution channels, plus cause you to lose the ability to incrementally improve your application, lose control and profit, and add to the device fragmentation problem. Plenty of short-term opportunity exists with native apps, but not without great personal risk, not to mention damaging the long-term viability of the mobile content market.

The most interesting case for mobile web apps is actually composed of the reasons stated earlier. If those are the only reasons that you need to create a native app, then what happens if you take away those obstacles from the mobile browser? Something like that is being done by Palm's webOS. They have created an entire mobile operating system built on WebKit, turning the phone into a web browser. Your "native applications" are just web applications.

Another innovation is the PhoneGap project, an open source effort that allows you to create native applications for iPhone, Android, and BlackBerry devices, exposing many device features like location and filesystem access to your web app. These applications can be distributed and sold in device marketplaces, but they share the same code and design. And because it is a web app, we can make a less capable version of our app available for free to lower-end mobile browsers. Build once, and deploy everywhere.

For those who have spent some time in mobile development in the past, you might have noticed the omission of "If you want to create a rich experience" from the reasons why to make a native app. Although there are certainly plenty of devices out there where this still might be the case, you can now create incredibly rich interfaces with mobile web apps. Not only can they be just as compelling as native apps, but they can work across multiple device frameworks with no alterations in code.

The rate of innovation for creating mobile web apps across every mobile device maker is at its highest level in years, but more important than that, for the first time device makers are all working toward achieving the exact same standards, which just happen to be the same standards as the desktop web. In addition, the devices that either lead in mobile web app innovation or support third-party browsers that do, are becoming the top-selling devices in multiple markets.

So, instead of asking yourself, "When to make a mobile web app?" I challenge you to start asking yourself "When *not* to make a mobile web app?"

## CHAPTER 10

Mobile 2.0

You've probably heard the term "Web 2.0." Although it's a commonly used term, most people you ask in the web business can't tell you what it means. To put it simply, it is just a label for the second generation of the web industry. But more importantly, it is meant to denote a change in that what we now believe in and stand for, which is not as it was before. The suffix 2.0 in technology implies that a product is new and improved, reinvented to be better and maybe even more relevant.

In the 1990s, we saw the adoption of the first generation of web technology, allowing businesses to create websites focused on the consumer market at that time. The "dot-com boom" was not about the Web; it was actually more of a boom in networking computers via the Internet, mostly driven by desire to use email as a communication and productivity tool.

In the early 2000s, the Web found its own voice. Though the technology had only incrementally evolved, the production costs dropped at the same time the market increased, creating exciting new opportunities to increase communications and productivity. Personal publishing became simple and easy to do, allowing more people to share more information and new ideas built on common problems to gain wider traction and adoption. The result was a fundamental change in how we create products for the Web, including everything from how we code, to how we design, even down to how we do business.

A few years ago, the mobile community started to discuss the idea of "Mobile 2.0," borrowing from many of the same principles behind Web 2.0. Each of these principles serves to transform the Web into a more agile and user-centered medium for delivering information to the masses. Mobile development, under the bottlenecks of device fragmentation and operator control, is sorely in need of a little reinvention as well.

Following is a recap of the original seven principles of Web 2.0:

The Web as a platform
> For the mobile context, this means "write once, deploy everywhere," moving away from the costly native applications deployed over multiple frameworks and networks.

Harnessing collective intelligence
> This isn't something the mobile community has done much of, but projects like WURFL—an open source repository of device profiles provided by the community—is exactly what mobile needs more of.

Data is the next Intel inside
> Mobile takes this principle several steps further. It can include the data we seek, the data we create, and the data about or around our physical locations.

End of the software release cycle
> Long development and testing cycles heavily weigh on mobile projects, decreasing all hopes of profitability. Shorter agile cycles are needed to make mobile development work as a business. Releasing for one device, iterating, improving, and then releasing for another is a great way to ensure profitability in mobile.

Lightweight programming models
> Because mobile technology is practically built on enterprise Java, the notion of using lightweight models is often viewed with some skepticism. But decreasing the programming overhead required means more innovation occurs faster.

Software above the level of a single device
> This effectively means that software isn't just about computers anymore. We need to approach new software as though the user will demand it work in multiple contexts, from mobile phones to portable gaming consoles and e-book readers.

Rich user experiences
> A great and rich user experience helps people spend less time with the software and more time living their lives. Mobile design is about enabling users to live their lives better.

Although the mobile industry has been through many more evolutions than just two, the concepts behind Web 2.0 are some of the most important ideas in not just mobile technology, but the Web as a whole.

What Is Mobile 2.0?

In autumn 2006, I was asked to help design and build a website for the first Mobile 2.0 conference, happening a few days before O'Reilly's Web 2.0 Summit. The event was put together by several Mobile Monday organizers. Mobile Monday is a series of mobile social gatherings that happen all over the world on the first Monday of the month. The idea was to bring some of the greatest minds in the mobile field together in San Francisco to attend the Web 2.0 Summit in order to discuss the future of mobile development.

By the time the event arrived, I had become frustrated with the term Web 2.0. By this time, it had devolved from a great idea to jargon that people would casually toss about. Everyone wanted "Web 2.0," but no one understood what it meant. Needless to say, I was skeptical about anything labeled 2.0. So I went to the first Mobile 2.0, curious about exactly what my fellow speakers thought Mobile 2.0 actually meant.

By the end of the day, I had a loosely defined picture of what Mobile 2.0 was, and for the first time in a long time I was excited about the future of mobile development. I saw a groundswell from the people at the conferences to finally overcome the industry bureaucracy and technical problems that had gone unsolved for too long. I saw not only what mobile technology could be, but people willing to make it happen.

In an interesting twist of foreshadowing, the first Mobile 2.0 event occurred only a few months before the announcement of the first iPhone—a device that would come to symbolize many of the things we talked about that day.

I took away many principles about Mobile 2.0, and about the future of mobile, which still hold true today. Like in Web 2.0, these ideas cover the industry from principles to techniques. The principles of Mobile 2.0 discussed in the following sections try to define the direction mobile development is headed in, sometimes by calling out the problems the medium faces still to this day.

Mobile 2.0: The Convergence of the Web and Mobile

It is obvious that in the minds of many, Mobile 2.0 is the Web. At this point, the mobile web has always been viewed as a second-class citizen within the mobile ecosystem, for many reasons, as discussed later.

Mobile is already a medium, but the consensus is that by leveraging the power of the Web, integrating web services into the mobile medium is the future of mobile development. When the iPhone exploded onto the scene, it increased the usage of the mobile web by its users to levels never seen before. The spur of new mobile web apps created just for the iPhone doubled the number of mobile websites available in under a year.

If Web 2.0 taught us that the Web is the platform, then Mobile 2.0 tells us that mobile will be the primary context in which we leverage the Web in the future.

The Mobile Web Browser As the Next Killer App

If the future of mobile is the Web, then it only makes sense that the mobile web browser is the next killer app of mobile. Again, this is something we saw confirmed with WebKit in the iPhone and later in Android.

However, of particular concern is how device fragmentation factors into mobile browsers. For example, how can we expect developers to support more than 30 different mobile browsers? A fellow panelist from the Mozilla Minimo project offered a potential solution in consolidation—that we will see only a few mobile browsers in the future; specifically, browsers built on Mozilla, Opera, Internet Explorer, and WebKit technologies. At the time, I thought that prediction was too focused on smartphones, but in the years since, the line between smartphone and feature phone seems to be going away, so this prediction is fairly accurate.

But the single biggest challenge in mobile remains device fragmentation. The mobile browser enables us to penetrate the problem by not having our content locked so specifically to the device abilities, screen size, and form factor, but device fragmentation still causes old, outdated browsers to remain in the market long after they should be put out to pasture.

What appears to be solving browser fragmentation is actually the iPhone. The Mobile Safari browser included with the iPhone provided such an excellent web experience on a mobile device that it drove use of the mobile web to huge levels, which means big profits for the operators. This also means that the mobile web is no longer a second-class citizen. In the post-iPhone market, all new devices are judged by the quality of their mobile web browser. Operators know it and therefore are demanding better browsers from device makers and browser makers.

Mobile Web Applications Are the Future

Creating mobile web applications instead of mobile software applications has remained an area of significant motivation and interest. The mobile community is looking at the Web 2.0 revolution for inspiration, being able to create products and get them to market quickly and at little cost. They see the success of small iterative development cycles and want to apply this to mobile development, something that is not that feasible in the traditional mobile ecosystem.

Developers have been keen for years to shift away from the costly mobile applications that are difficult to publish through the mobile service provider, require massive testing cycles and costly porting to multiple devices, and can easily miss the mark with users after loads of money have been dumped into them.

The iPhone App Store and the other mobile device marketplaces have made it far easier to publish and sell, but developers still have to face difficult approval processes, dealing with operator and device maker terms and porting challenges.

Mobile software has two fundamental problems that mobile web applications solve. The first is forcing users through a single marketplace. We know from years of this model that an app sold through a marketplace can earn huge profits if promoted correctly. Being *promoted correctly* is the key phrase. What gets promoted and why is a nebulous process with no guarantees. One thing is certainly true: the companies that know how to work the system are the ones that get the big prizes, making it increasingly hard for the small developer to see any kind of success. But the mobile web provides any size of developer with the ability to promote and distribute their app on their terms, building a relationship directly with their customers and not by proxy.

The second problem is the ability to update your application. It is certainly possible on modern marketplaces like the App Store, but we are still years from that being the norm. Mobile web apps enable you to make sure that you never ship a broken app, or if your app breaks in the future due to a new device, to be able to fix it the same day the device hits the street. This flexibility isn't possible in the downloaded app market.

JavaScript Is the Next Frontier

If you are going to provide mobile web applications, you have to have a mobile web browser that supports Ajax, or, as it is technically known, XMLHttpRequest. It makes a lot of those cool interactions in your web browser work via the capability to load content asynchronously into your browser view.

But it isn't just Ajax; it is JavaScript, a web technology that has largely been ignored with most mobile web browsers. Ajax is great, but just being able to do a little show/hide or change a style after you click or touch it goes a long way toward improving the user experience.

This is probably where mobile web browsers fall behind desktop browsers the most. Because they both support XHTML and CSS relatively well, JavaScript has been a no-go in mobile for years. In order for mobile web apps to rival native applications, you have to support some JavaScript.

Modern mobile browsers have made much progress over the last few years, but there is still plenty of work to be done. For example, accessing the device capabilities like the phone book or filesystem with JavaScript doesn't work in a consistent way. These problems still need to be solved in order to truly reap the benefits of the Web.

Rich interactions kill battery life

JavaScript and Ajax have been ignored because using an Ajax-based web application on your phone can drain your battery at a rate of four to five times your normal power consumption. I've heard a number of reasons for why this happens from mobile hardware guys much smarter than myself, but to summarize, the two most prevalent are:

- JavaScript consumes more processor power and therefore more battery life.
- Ajax apps fetch more data from the network, meaning more use of the radio and more battery life.

Unless you are in the habit of carrying around a bunch of extra batteries, expect to charge your phone every hour or two as a penalty for using the modern mobile web.

Apple and the open source WebKit browser have made huge strides by releasing a JavaScript engine that is incredibly efficient on mobile devices, though the other big mobile browser technologies aren't far behind. This problem is going away quickly as the mobile browsers get better, batteries improve efficiency, and devices get more powerful.

The Mobile User Experience Is Awful

Traditionally, the user experience available in the mobile web has been like using a website from 1995: mostly text-based, difficult to use, and ugly as sin. This isn't to say that the user experience of mobile applications has been much better, but it used to be that if you wanted a good experience, you built a native app.

Descriptions within the industry range from the honest "the mobile user experience is utterly horrid," to the sales pitch of "look at these cool things you can do," to the optimistic "the mobile user experience is the future!" These polar attitudes toward the mobile user experience are somewhat ironic, given that the mobile user experience was largely ignored for close to a decade. People in mobile treat the user experience like a chicken-and-egg scenario: bad input/output of the user experience prevents adoption, but designing a shiny user experience with bells and whistles will bring them in droves.

Device APIs usually force you to use their models of user experience, meaning that you have to work in the constraints of the API. This is good for creating consistent experiences for that platform, but these experiences don't translate to others. For example, would you take an iPhone app design and put it on an Android device? The user experience for these devices is similar but still remains different.

The beauty of the Web, literally, is that you can design whatever experience you want, for better or worse. You are in control of the presentation and can establish your own visual metaphors. The problem has been that traditionally complex (which often equates to good) user experiences haven't been possible on mobile devices. Modern mobile web browsers, as they come closer to their desktop counterparts, remove this distinction, giving us the same canvas on mobile devices that we have for the desktop. This means that creating mobile experiences just got a whole lot easier. It also means we can have a consistent user experience across multiple mediums.

Mobile Widgets Are the Next Big Thing

At many Mobile 2.0 events, I've heard a lot of buzz about mobile widgets, though no one can tell me how mobile widgets would define a mobile widget, or how they are different from mobile web apps. The consensus seems to be that the solution for the challenges with the mobile web is to create a series of "small webs" targeted at a specific user or task. Though I couldn't figure out the problem being solved with these widgets, I had to admit that they looked pretty cool.

Don't get me wrong. I believe that the concept of small network-enabled applications is very promising, but the mobile industry tends to take promising ideas like this, inflate expectations to unsustainable levels, then abandon them at the first sign of trouble or sacrifice them for the next big thing, whichever happens first.

The mobile web is here: it works, but it just needs some love. It is the long bet. I believe we should get that sorted out before we try to add widgets on top of it.

Carrier Is the New "C" Word

I noticed a strong tendency over the years for people in the mobile industry to avoid uttering the word "carrier." Even the more European equivalent term "operator" seems to be on the decline. To give you an example of how much carriers are hated: you can have entire conversations with people who work for the big operators and even they will avoid using the term in conversation.

It is almost like when you get a bunch of mobile experts in a room for a day: they want to pretend that operators don't exist at least for one day. Maybe they prefer to see a future with no mobile service providers at all. I think more likely the case is that the industry has finally figured out that very few can make a profit when your business relies on carriers. Though the "C" word isn't uttered often, it still is the 800-pound gorilla in the room.

It is clear that one of the key drivers of Mobile 2.0 and the focus on the mobile web is to find a way to build a business that doesn't rely on carrier control.

Mobile Needs to Check Its Ego

For years, I've sat on the line between the mobile community and the web community. They have treated each other almost like rivals. I've been frustrated with both sides, but it is the mobile camp that needs to check their egos at the door and get into the game, before they learn that all the rules have changed.

On the mobile side, you have some incredibly intelligent people who have been innovating amazing products under insane constraints for years. On the web side, you have creative amateurs who have helped build a community and ecosystem out of passion and an openness to share information.

The web guys want to get into the game and move the medium forward, partly out of desire open up a new market for themselves, but mostly out of passion for all things interactive. But, to the mobile community, they are seen as a threat to expertise. On the other hand, to the web community, the mobile guys come off as overly protective, territorial, selfish, and often snobbish or egotistical, effectively saying, "Go away."

Don't get me wrong—I think that the mobile guys are very smart and great people; some of them I consider to be close friends. They have to deal with really hard problems that would make a web professional give up to go serve coffee. But I'm a very patient and tolerant person, and I have to admit that these same people are some of the most difficult people I've ever worked with.

For example, for years I wondered why I, a designer with mobile experience, have been asked to speak at so many conferences or write so many articles. I would think to myself,

"I can think of 10 guys who know a lot more about mobile than I do. They should be here, not me." But then this guy who came up to me after one of my workshops said, "I've been in mobile for a long time, and have been to a lot of mobile events, but no one has ever spoken about mobile so plainly." I thanked him for the wonderful comment and replied, "Well, that's the problem with mobile, isn't it?"

My point is that unless the mobile community comes together with the web community by sharing information, experience, and guidance, one day they will find that their experience has become obsolete. In return, the web guys will share their enthusiasm, willingness to learn, and passion that many in mobile development have forgot.

It's that one principle of Web 2.0 that the mobile community has left out: harnessing collective intelligence. The Web and the mobile community are reaching a point where the two worlds can no longer afford not to be working together, sharing what they know and harnessing the collective intelligence of both media.

We Are Creators, Not Consumers

The final principle of Mobile 2.0 is recognizing that we are in a new age of consumerism. Yesterday's consumer does not look anything like today's consumer. The people of today's market don't view themselves as consumers, but rather as creators. But before we get into that, let's back up for a minute.

The web is about content. Sure, there are programming languages, APIs, and other technical underpinnings, but what do you do when you open a web browser? You read. Our primary task online is to read, to gain information. During the early days of the Web, it took tools and know-how in order to publish to the Web. But early in the Web 2.0 evolution, we saw a rise in tools that allowed us to publish to the Web easily, giving individuals a voice online, with a massive audience.

This democratization of the Web took many forms that some call "social media," like blogging, social networks, media sharing, microblogging, and lifestreams. Although social media may have many facets, they all share the same goal: to empower normal, everyday people to become creators and publishers of content. It started with the written word, then music, then photos, and more recently video was added. Entire markets have been created to provide today's consumer with gadgets, software, and web services to record and publish content so that we can share it with our friends and loved ones.

At the center of this revolution in publishing is the mobile device. As networked portable devices become more powerful, allowing us to capture, record, and share content in the moment, we are able to add a new kind of context to content—the likes of which we haven't seen since satellite television. Now you can share any moment with any group of people in real time. Think about how powerful a concept that is! It could change entire cultures.

Tony Fish, coauthor of *Mobile Web 2.0* (futuretext), says:

> When everyone has the tools to create content, in addition to zero-cost publishing, we do not consume content, we create it.[*]

In the early days of the Web, I marveled at how a networked population might change our society forever. Now I realize that the change occurs wherever the device is, the context it is within. The early "Web 1.0" days clearly changed how business is done, because businesses are the primary consumer of desktop computers. It probably is no coincidence that Web 2.0 occurred around the same time that laptop computers became affordable for the average person, making the Web a more personal medium.

With Mobile 2.0, the personal relevance of the content matches how personal the device is and how personally it applies to our everyday situations or our context. I see now that this is the time and medium that delivers on that initial promise of the Web: to change society forever.

---

[*] Sources: *http://communities-dominate.blogs.com/brands/2007/02/mobile_the_7th_.html* and *Mobile as 7th of the Mass Media*, by Tomi Ahonen (futuretext).

## CHAPTER 11

Mobile Web Development

In this chapter, I discuss the foundational principles and techniques for creating designs for multiple mobile devices. As discussed earlier, the mobile web is the only ubiquitous platform for delivering content to mobile devices. This makes it an incredibly powerful platform in terms of its reach, but like most things in mobile, every blessing comes with a curse. The price for ubiquity is that not all designs are rendered the same.

There are two basic approaches to creating cross-platform mobile designs: content adaptation, in which you create many versions of your site, or no content adaptation, in which you don't. In either case, you need to have a firm understanding of how to design content for the mobile context. Though I won't discuss the pros and cons of adapting for devices until Chapter 13, for now it is important to understand that the rendering of many of the standards and techniques I discuss in this chapter are measured on a sliding scale of consistency across multiple devices.

Now this doesn't mean that mobile standards are not standardized. Many coming from the desktop web world mistakenly assume that because a mobile browser renders elements inconsistently, there is no web standards support within the mobile web—this isn't exactly the case. The mobile web actually has been very standardized, with defined specifications—in some cases, with standards older than the specifications we use on the Web. The problem is the technical constraints of devices and the inability to update the browser technology that ships with devices. Together with low consumption of the mobile web, the problem has been somewhat ignored for years.

That all changed, of course, after the introduction of the iPhone. I will talk about some of the changes that are occurring in the mobile web and some of the techniques specific to creating iPhone web apps in following chapters, but for now I want to introduce you to the foundation of what makes the iPhone and other advanced mobile browsers successful: creating mobile experiences under extreme constraints. It doesn't matter what mobile devices you plan to support; they will always have the same challenge. These principles and techniques were designed and honed over the years to address exactly that.

Web Standards

The first question that probably pops into your head (as it certainly does in mine as I write this) is why do we need to have a chapter specifically about mobile web development at all? Isn't that what the concept of "web standards" is all about? To separate content from presentation in order to create a ubiquitous experience that can be rendered on any device or in any viewing context? In a word, yes, but unfortunately it isn't that cut-and-dry.

The phrase "web standards" is often used to support arguments, like "you should be supporting web standards because...." But what does it really mean? By my definition, web standards is just an easy way to say "a web page based on the XHTML 1.0 and CSS 2.1 specification, coded in such a way that the majority of presentation elements are omitted from the XHTML markup and defined in the CSS instead." Now it is typically inferred that in doing so, your markup is immediately made more usable in other viewing contexts—for instance, search engines and accessibility devices such as screen readers and mobile devices—but that is not necessarily the case, as any search engine optimization (SEO), accessibility, and mobile web expert, respectively, will quickly tell you.

A common misconception that web designers and developers have about XHTML is that it should be treated like XML, as a structural data source rather than as a tool to mark up content, like XHTML and its forefather, SGML, were designed to be. In fact, XHTML is not the most common language used for most large-scale mobile deployments; instead XSLT, or Extensible Stylesheet Language Transformations, is the preferred choice. In this method, content is defined as XML and then XSLT is used, along with multiple markup languages like HTML, XHTML, WML, XHTML Basic, XHTML-MP, and so on, to provide the proper rendering markup for the viewing context.

Though mobile device browsers are certainly becoming better at rendering what we refer to as "standards" consistently, that isn't to say that XHTML 1.0 and CSS 2.1 is, or even should be, the one and only means to display content on the Web to mobile devices. The beauty of the mobile web is that it can take form in so many shapes, creating highly useful tools in many contexts. Trying to rigidly adhere to web standards dogma not only hurts the mobile web, but the desktop web as a whole, setting manufactured constraints over how the Web can evolve.

As the youngest of six sons, I can attest that my parents used very different methods to rear each of us. I can't help but look at the web standards community in the same way in which my parents must have looked at my brothers and myself. Each of us had our path, our own direction. Trying to apply the same rules, the same method of parenting, to each of us just didn't work. They had to adapt to each of our personalities and foster a safe environment for each of us to learn and grow at our own pace, to discover our own potential, while still teaching us the beliefs and guiding principles they wanted us to share.

Following my parents' example, I tend to look at the desktop and mobile web as siblings. They have a lot in common, but they are not the same—nor will they ever be. They each have their own path and their own direction, and we cannot always expect the same guidance and methods to work exactly the same on each. The best we can do is foster each of them to reach their greatest potential.

Designing for Multiple Mobile Browsers

Designing and developing for multiple mobile browsers simultaneously is a challenge, but not an impossibility. It requires looking at your designs and code from many contexts, and being able to visualize how your designs will be rendered on a variety of devices in your head, as you lay down code.

For example, you are creating the markup and styles for a desktop site that has to support Internet Explorer 6, which has very quirky support for web standards techniques. You would employ a different technique to express your design, one that you know is proven to be compatible on most browsers, than if you were just going to support the latest browsers. In mobile development, we have might have 10 Internet Explorer 6s, so we have to think of our design in layers of degradation, which just happens to be the definition of progressive enhancement.

Progressive Enhancement

Progressive enhancement is the practice of using web techniques in a layered fashion to allow anyone with any web browser to access your content, regardless of its capabilities. This means that you are creating not just one ideal experience, but multiple less-ideal experiences, depending on who views the content, also called *graceful degradation*. To illustrate this, take a look at Figure 11-1.

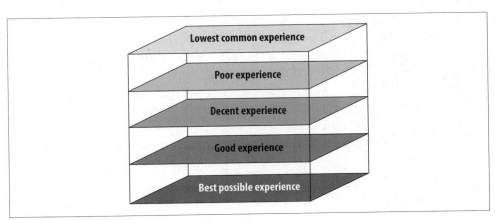

Figure 11-1. Using progressive enhancement to layer support

Our unstyled experience or markup viewed in its rendered form is at the center. We can assume that all devices can show at least this level of presentation. However, in order to make sense of it we need to ensure our markup is written semantically, in other words, ensure that it can remain useful without a presentation. Next, we add basic styling techniques for our lowest common denominator devices that support both our markup plus very basic styling techniques. We continue to add layers until we reach the best possible experience on the outside.

This can be done with or without content adaptation, although my advice is to always start your project without. It's far faster and easier to start with one code base that gracefully degrades than immediately jumping into creating multiple versions of your site.

It should be noted that you can add a desktop layer to your site using progressive enhancement. There is absolutely nothing wrong with this approach, because you built the desktop layer on top of a layer designed for the mobile context. In fact, this is something I did with the original Mobile 2.0 event site (Figure 11-2), creating a useful mobile experience, then creating multiple layers for different viewing contexts, and ending with the desktop.

The problems arise when you go the other way around, starting with the desktop and trying to create a useful mobile experience from it. Nine times out of ten, you end up with an awkward, barely useful version of your site that is both slow and expensive to load.

Mobile progressive enhancement techniques

The following tips are tricks I've been using since the early days of the mobile web. Regardless of how advanced browsers have become, I still employ these five techniques for every mobile project I do:

- Always code your markup semantically. I like to keep a live preview window open so that I can see the rendered page as I code. This way I can ensure that the page is always usable, even with no stylesheet in place.
- Have a device plan. Know which device classes you intend to support before you start to code. This will influence how you code your pages.
- Have both your lowest common denominator and high-end device designs before you begin to code. Try to visualize a way to create both versions from one code base.
- Test on different mobile devices from the beginning to the end to ensure that your incremental work will display correctly in the intended devices.
- If you plan to add a desktop layer, always create the mobile version first.

Figure 11-2. The Mobile 2.0 site, using progressive enhancement techniques, producing both a desktop version and a simple mobile site

DIAL

Seeing a growing need to produce a standard means of markup for multiple contexts, the W3C has begun to develop a specification for a device-independent authoring language referred to as DIAL. DIAL is an XML language profile of XHTML2, shedding its SGML roots and becoming more of a means of delivering different content to different devices by using machine-readable constructs to define conditions.

Take the W3C's example:

```
<!-- query the requesting device's browser resolution in dpi and store in a convenience
variable-->
<sel:variable name="res" value="di-cssmq-resolution('dpi')"/>
<sel:select>
 <sel:when expr="$res &gt; 500">
  <object sel:selid="Cornish Yarg" src="yarg_hi.jpg"/>
```

```
    </sel:when>
    <sel:when expr="$res &gt; 200>"
     <object sel:selid="Cornish Yarg-id001" src="yarg_mid.jpg"/>
    </sel:when>
    <sel:otherwise>
     <object sel:selid="Cornish Yarg-id001" src="yarg_low.gif"/>
    </sel:otherwise>
   </sel:select>
```

In order for DIAL to work, it must pass through at least one DIAL processor in order to render the desired view for the desired device. In other words, we are talking about content adaptation, but at least we are talking about a standardized means of content adaptation. In fact, several content adaptation server vendors already favor the DIAL specification over the proprietary XSLT techniques, even though the specification is still a working draft.

Though I think this is promising technology for mobile sites planning to employ some means of content adaptation, I'm a bit skeptical. As we've been seeing the future of markup standards lean toward evolutionary and not revolutionary (as in the case of HTML5 versus XHTML2, which I discuss more in the following chapter), I have a feeling it will still be some time until you are coding your site in DIAL. When desktop, mobile, and other browsers all start supporting DIAL as a client-side processor, I think it will start to take off. But until then, most likely it will largely remain solely a content adaptation technique.

Designing for Multiple Displays

Next on my list of things to cover is supporting multiple mobile displays. It is both a design and development dilemma, but if you ask any mobile designer, it is quite a painful headache. When trying to design and develop a mobile experience, you have to remember that your design might be viewed on a small 120-pixel screen, common on most lower-end phones, or on a 320-pixel screen common to most smartphones. Depending on the type of devices you plan to support, it is entirely possible that it could be viewed on larger screens; many smartphones now can be rotated to landscape mode, increasing the view to 480 pixels. To cover all bases, I tend to look at any screen under 760 pixels, the maximum viewable size on an 800×600 desktop resolution, as a possible mobile device.

Fixed versus fluid designs

Traditionally, in the mobile industry developers have opted for fixed-width designs over fluid, or percentage-based, stretchable designs. Fixed-width has provided slightly more reliable rendering across devices in the past; however, the problem with fixed-width design is that it might limit the viewable content when viewed on larger screens. We are seeing a trend toward larger screens as well as the inclusion of orientation switching in device browsers. Therefore, I recommend that you make all designs fluid by default, unless you are targeting a device that you specifically know has issues rendering a fluid design.

Single-column versus multiple-column layouts

As mobile screens get larger, does that mean we can start to use multicolumn layouts? For example, should you use sidebars or vertical navigation areas? I honestly don't know. I will leave it up to you to determine whether it is the best presentation for your users. I've used a user-collapsible vertical navigation area for mobile products in the past, and my users loved that they could toggle between the two. However, the vast majority of mobile designs opt for a simple single-column layout.

There are historical, technical reasons why you might not want to employ a multicolumn layout on a lower-end device: poor support for positioning or floats, not to mention that these devices typically have smaller screens where two columns would make the page cluttered, are just a couple of examples. Also, if the user has a device with a directional pad, or D-pad, then multiple columns would create an awkward scrolling pattern.

My rule of thumb is actually to consider whether you're designing for touch. If your target devices support touch, then it is OK to use multiple columns, but tread forward carefully and don't make the page difficult to use. If the device does not support touch, then opt for a single-column design.

Device Plans

Developing a mobile product means having a device plan at the very start. Effectively, you've defined each of your progressive enhancement layers, determining what will be that center, common experience and what layers you intend to support. Unfortunately, I've seen it too many times: trouble at the late stages of developing a mobile project because there was no defined device plan. The goal is to get to the test stage without any surprises. Sure, some things might not render properly, but you should at least have an idea of why the product is broken. Not having a decent device plan can mean getting into the test phase and having mind-boggling bugs that you didn't expect—and worse, no idea how to fix them.

I think of my device plan as a passage from Sun Tzu's *The Art of War* (Delacorte Press):

> If you know both yourself and your enemy, you can come out of hundreds of battles without danger.

In this case, the invading Mongols trying to breach your wall are those hundreds of devices. If you know the weakness of each of the devices you plan to support, along with what you will be able to safely accomplish, then you can have many successful mobile projects.

But getting a handle on hundreds of mobile devices is no small feat. Therefore, I use a system of classifying browsers into five broad groups. Doing this gives us a feature set and requirements to work within and support.

The Device Matrix

Table 11-1 provides a listing of popular browsers and their assigned classes, starting with A, the highest grade, to be considered on par with desktop browsers, and ending with F, the lowest possible grade.

Table 11-1. The device matrix

Class	Markup	CSS	JavaScript
Class A	XHTML, XHTML-MP, HTML5	CSS2, CSS3	Great, includes DHTML, Ajax
Class B	XHTML, XHTML-MP	CSS2 (Decent)	Limited, some DHTML
Class C	XHTML, XHTML-MP	CSS2 (Limited)	Limited
Class D	XHTML-MP	CSS2 (Basic)	None
Class F	XHTML-MP, WML	None	None

Class A mobile browsers

Some of the characteristics of a Class A mobile browser are:

- Excellent XHTML 1.0 support
- Good HTML5 support; specifically, the canvas element and offline storage
- Excellent CSS support, including most of CSS Level 2.1 (scores 90 percent or higher on the ACID2 test) and the majority of CSS Level 3 (scores 75 percent or higher on the ACID3 test)
- Support for web standards layouts, including absolute positioning, floats, and complex CSS-based layouts
- Support for image replacement techniques
- Excellent JavaScript support
- Ability to toggle the display property
- Support for DOM events, including Ajax
- Considered comparable to a "desktop-grade" browser

Class B mobile browsers

Some of the characteristics of a Class B mobile browser are:

- Excellent XHTML 1.0 support
- Good CSS Level 2.1 support (scores 75 percent or higher on the ACID2 test)
- Padding, border, and margin properties are correctly applied

- Can reliably apply colors to links, text, and background
- Supports image replacement techniques
- Minimum screen width: 164 pixels
- Can support complex tables—not necessarily nested tables—up to four cells in a row
- Setting a font size of 10 pixels or more produces readable text
- Has limited JavaScript support, being at least able to toggle the display property

Class C mobile browsers

Some of the characteristics of a Class C mobile browser are:

- Good XHTML 1.0 support
- Limited CSS Level 2.1 support (scores 50 percent or higher on the ACID2 test)
- Limited or no JavaScript support

Class D mobile browsers

Some of the characteristics of a Class D mobile browser are:

- Basic XHTML
- Limited CSS support (CSS Level 1, or does not recognize cascading)
- Minimum screen width: 120 pixels
- Hyperlinks may not be colorable by CSS
- Basic table support: 2×2 or more
- colspan and rowspan may not be supported
- "Width" expressed as a percentage may be unreliable
- No JavaScript support

Class F mobile browsers

Some of the characteristics of a Class F mobile browser are:

- No (or very unreliable) CSS support
- Poor table support or none at all
- Basic forms: text field, select option, submit button
- May not be able to support input mask on fields
- No JavaScript support

Markup

Markup is used to make content readable by mobile browsers. Normally when we think of markup, we think of HTML, or Hypertext Markup Language, the language of the Web. In mobile, we use slightly different markup languages, depending on the scope and size of your project or on your target devices.

Back in 1996, HDML (Handheld Device Markup Language) was created by Openwave, one of the early pioneers of the mobile web. It was quickly followed by Nokia's TTML (Tagged Text Markup Language) and other proprietary markup languages, before the WAP (Wireless Application Protocol) Forum created the WML (Wireless Markup Language) specification in 1998, which was included in the WAP 1.0 specification.

Many people coming from the Web assume the mobile web works in the same way as and lives on the same protocols the desktop web does, which isn't the case. WAP is the stack in which the mobile web lives, but it is a protocol unto itself. Saying that WAP and the Web are synonymous would be like saying that men and women are the same. Yes, they are both humans, and technically how they function is very similar to one another, but there are some fundamental and obvious differences. Mobile devices are much more like their desktop computer counterparts today, but there are still many remnants of WAP within the current operator-delivered mobile web.

WML looks more like XML than HTML, using a strict format to mark up content, organized into cards, or pages; therefore an entire WML site would be referred to as a "deck," a term still used today to describe a mobile website. As devices and networks matured, so did the markup, moving away from the XML syntax and closer to the HTML syntax, first with the cHTML used with NTT DoCoMo's i-mode phones, then iHTML. It was followed by XHTML and XHTML Basic, which evolved into XHTML-MP (Extensible Hypertext Markup Language—Mobile Profile), a modularization of the XHTML that we all know and love, but for mobile devices. The Open Mobile Alliance (OMA) has defined XHTML-MP as the primary language of the WAP 2.0 protocol, and it has been commonly used in phones since 2002.

XHTML-MP Overview

XHTML-MP is a modularization of XHTML Basic. XHTML Basic is a subset of XHTML. In other words, if you know XHTML, chances are good that you will be comfortable with XHTML-MP. Now that XHTML Basic and XHTML-MP are virtually indistinguishable, and thanks to both standards being a subset of XHTML—the standard language of the Web—the average web developer does not need to understand a language for each medium.

XHTML-MP has evolved to become the predominant language for the mobile web, and can safely be assumed to be used in devices manufactured since 2002. Operators used to require on-deck mobile websites to be coded in both WML and XHTML-MP, using WML as the fallback. Luckily, those days have long since passed. However, anyone considering the creation of mobile web content for emerging markets like African countries had better brush up on their WML skills, as WML can still be found in pockets that use largely recycled phones.

XHTML versus XHTML Basic versus XHTML-MP

Some of you may be scratching your heads at this point. No doubt you've never seen the letter X used so many times in one page. But what does this mean for you? Should you code your mobile site in XHTML, which we use on the Web? Or maybe XHTML Basic, which is on the W3C site? What about XHTML-MP, which this Open Mobile Alliance you've never heard about recommends?

Though they are almost indistinguishable from each other, it really just depends on the devices you plan to support. You are certainly safe with XHTML-MP on any mobile device. Although more advanced devices like the iPhone prefer XHTML over XHTML-MP, if my device plan calls for consistency on lots of lower-end devices, there is no doubt in my mind that I should use XHTML-MP. And if I plan to have the iPhone being at the high end, or to add a desktop layer, I use XHTML.

Because XHTML-MP is a modularization of XHTML, many devices—even low-end devices—will support an XHTML document. It just might be a bit more flaky. So ask yourself where your users are most likely to be and go from there.

Document Structure

The following are guidelines, recommendations, and best practices to structure your XHTML-MP documents appropriately. Different classes of browsers treat each of these best practices differently, so use the following list to determine how closely you should adhere to them:

- Class A browsers: Recommendations, not mandatory
- Class B browsers: Best practices, should reduce inconsistencies
- Class C browsers: Strongly recommend, veering from will increase inconsistencies
- Class D browsers: Required, should adhere closely
- Class F browsers: Required, but may still produce inconsistencies

Doctypes

Applying the XHTML-MP doctype tells mobile browsers how to render the content. Using the incorrect markup type specified by the doctype can cause content to render erratically or incorrectly. Defining the XHTML-MP doctype will provide the most reliable rendering of the page across the widest range of mobile devices.

For Class B and lower devices, use the following doctype:

```
<!DOCTYPE html PUBLIC "-//WAPFORUM//DTD XHTML Mobile 1.0//EN"
"http://www.wapforum.org/DTD/xhtml-mobile12.dtd">
```

However, for Class A browsers, use your favorite flavor of XHTML doctype. I like XHTML 1.0 Transitional:

```
<!DOCTYPE html PUBLIC "-//W3C//DTD XHTML 1.0 Transitional//EN" "http://www.w3.org/
TR/xhtml1/DTD/xhtml1-transitional.dtd">
```

Character encoding

Correct character encoding is essential to making sure that pages render correctly on devices. Different document types require different character encodings. XML documents should always have a UTF-8 character set; documents served as MIME type text/html should use ISO 8859-1. The following line shows how to set encoding correctly in an XML document:

```
<? xml version="1.0" encoding="UTF-8" ?>
```

Specify the correct character encoding for your pages; otherwise, your page may display strange characters. The recommendation is to use UTF-8 encoding for maximum compatibility. Pages delivered with the `text/html` MIME type are assumed, by default, to be encoded with ISO-8859-1. If you are using `text/html` and delivering UTF-8, set the HTTP Content-Type header to state this. If you develop web pages in a Windows environment, note that the default character encoding is often Windows CP 1252—which is similar but not identical to ISO-8859-1.

MIME types

Servers sending MIME types provide important information to browsers on how to treat a document. Sending incorrect MIME types with a document may cause the browser to incorrectly interpret and fail to render the document. For XHTML-MP, the recommended MIME type is `application/vnd.wap.xhtml+xml`. Unlike HTML, XHTML-MP shouldn't serve text/html. Administrators often set up web servers correctly for common document types such as HTML and CSS, but not for XHTML-MP.

Page titles

Page titles surrounded by the `<title>` element are an important and frequently overlooked page element. Good titles increase the findability and usability of web pages. Add a short descriptive page title for easy identification, but remember that the mobile device may truncate the title. Most devices use the page title as a default label for bookmarks, so a title helps the user identify content.

It is common to use only the site name as the title, but this doesn't help users as much as other approaches. The title of the document should consist of the primary title, optionally followed with your site name, as shown in the following example:

```
<title>Description of Page Content | Site Name</title>
```

Search engines primarily use page titles to identify content. Also, think about how users will search for your content on search engines while naming your pages.

Use of stylesheets

Many mobile browsers prioritize markup before presentation, loading stylesheets and images last. This sometimes causes markup to appear with styles briefly while the external stylesheet loads, known as the "screen flash." You can avoid this by adding styles to the document head instead of using an external stylesheet. However, in doing so, you lose the ability to centrally manage your styles.

My recommendation is to always use external stylesheets, separating your markup and presentation and decreasing the overall page size.

Objects and scripts

Most mobile devices don't support embedded objects or scripts, and it's not possible for users to install plugins to provide support. Even where a device does support scripting, avoid using it unless you can't find another means to express your design. Though many modern browsers support scripting, you may want to skip it altogether so that you limit your power consumption and have fewer rendering inconsistencies to contend with.

If you must rely on either scripting or embedded objects, use them very sparingly and test often.

Auto refresh

Avoid creating periodically auto-refreshed pages unless you inform the user and provide a way to stop it, as shown in the following example:

```
<meta http-equiv="refresh" content="0" />
```

Redirects

Using markup to redirect pages increases the load time and cost as a result of downloading and processing another page. If you need to use redirects, configure the server to perform redirects using HTTP 3xx codes.

Caching

Using cached information sometimes reduces the need to reload resources such as images and stylesheets, thereby lowering download times and costs. By specifying cache information on your mobile pages, you lower the number of times devices download common resources. This especially helps resources like a stylesheet or logo, as shown in the following example:

```
<meta http-equiv="Cache-Control" content="max-age=300"/>
```

Not all devices support cache control, but caching is important for mobile devices due to the typically high network latencies experienced on mobile networks. Just be sure to wait until you complete development before adding cache information—otherwise, you won't see development changes, thanks to caching.

Minimal document structure

It is good practice for documents to indicate structure with headings and subheadings. Code in order and in a semantically correct fashion so that the code elements and the order in which they appear make sense without manipulating the presentation. The following is an example of coding semantically:

```
<h1>Top Level Heading</h1>
<h2>Second Level Heading</h2>
 <p>Paragraph Body</p>
<h3>Third Level Heading</h3>
 <p>Paragraph Body</p>
<h2>Second Level Heading</h2>
 <p>Paragraph Body</p>
<h3>Third Level Heading</h3>
 <p>Paragraph Body</p>
<h4>Fourth Level Heading</h4>
 <p>Paragraph Body</p>
```

Using structural markup, rather than formatting effects, makes it easier to modify content when it needs splitting into several pages. Furthermore, structural markup potentially facilitates access to the sections of the document that a user wants. Use headings in accordance with the specification whenever applying them. For example, they should properly nest based on their level, as in the previous example.

Text Elements

If you are familiar with XHTML, then none of the following should be any revelation, but for good measure I want to make sure that each of the common text elements used in an XHTML document are included, and to describe how they can differ when rendered on a mobile device.

Headings

```
<hn>...</hn>
```

Headings h1 through h6 are supported in XHTML-MP. Your typical mobile page doesn't have more than two or three headings on the page, for a few reasons. First, an overly structured document starts to lose its context with smaller screens; in other words, it can be difficult for users to be able to understand the header relationship when they can see only a few lines of text at a time. Second, headers are often rendered only a few pixels larger than the default text size. In desktop web browsers, headers are often significantly larger than the default paragraph size, but mobile devices cannot take this liberty.

In traditional mobile web authoring, headings would be omitted entirely, opting for bolded text, followed by a break in order to render more consistently. I heavily discourage you from using this technique and to code your pages semantically. This will give you great flexibility in your progressive enhancement strategy later.

Paragraphs

```
<p>...</p>
```

The paragraph is the tag you will probably use the most. Each paragraph of text should be wrapped in the paragraph tag. The paragraph will apply default margins to the top and bottom on the element, which can be modified in the CSS.

Historically, paragraph tags would not be used due to the poor CSS support of devices; break tags would instead be used to create line breaks between paragraphs. Again, I strongly discourage you from employing this technique.

Quotations

```
<bq>...</bq>
```

The blockquote is used for quotations or comments, and is often used as a wrapper tag for one or more paragraph tags. Blockquotes often inherit margin around the entire element to give the appearance of being indented from the primary text, which may not be rendered consistently on all devices. I recommend making sure that you define the margins for blockquotes in your CSS instead of relying on inherited margins, therefore making it easier to debug later.

Phrase elements

```
em, strong, small, abbr, acronym, cite, dfn, code,
kbd, samp, var, del, ins
```

With the exception of em and strong, I encourage you to not use any phrase elements in your document markup, as they may not be fully supported on Class C or lower devices. Instead consider using <span class="*phrase*">…</span> and control your phrase presentation in your CSS.

Unordered lists

```
<ul><li>...</li></ul>
```

Unordered lists are a hallmark in web-standards-based design, often used for navigation lists and structuring nested content. Unordered lists are also useful in mobile web pages. For Class A browsers, unordered lists are just as useful in your designs as they are on desktop sites. For Class B and lower browsers, advanced styling of unordered lists, like displaying items horizontally, can sometimes create inconsistencies across browsers.

I recommend starting with simple styling and performing incremental tests on your targeted devices to ensure that your desired styling will display appropriately across multiple devices. Otherwise, just keep your styling very simple and restricted to margin, padding, and bullet type.

For Class B and lower browsers, I would avoid creating nested lists. They typically render as they should, but given the limited screen width, having multiple nested lists means that the inherited left margin can create an unreadable display.

Ordered lists

```
<ol><li>...</li></ol>
```

Ordered lists are not used as often as unordered lists in desktop sites. For mobile designs, I recommend using ordered lists for all your navigation lists that have fewer than 10 items. This allows you to associate an access key to the appropriate navigation item:

```
<ol> <li><a href="#item1" accesskey="1">Item 1</a></li> <li><a href="#item2"
accesskey="2">Item 2</a></li> <li><a href="#item3" accesskey="3">Item 3</a></li></ol>
```

Definition lists

```
<dl><dt>...</dt><dd>...</dd></dl>
```

A definition list is for lists that contain term and definition pairs, and is useful for creating repetitive lists where you simply need a title and do not wish to use a header; this is often used to denote a section. For Class A browsers, I use definition lists to structure forms, allowing me to place the form label either above or to the left of the form input or control. When targeting Class B and lower browsers, I avoid using definition lists altogether, given their less-than-trustworthy ability to do any advanced styling.

Structural elements

```
<div>...</div><span>...</span>
```

The div and span elements are just as critical to mobile web development as they are to desktop web development. The div is used to identify and label any block-level division of text or content; this could be a line of text, or it could be an entire page, whereas the span tag is used to identify a grouping of inline elements and is often used within a block-level element.

The div and span are essential for adopting a progressive enhancement strategy, allowing you to conditionally define what content is seen for what devices. For example, we can show only the content with the class lcd to our lower-end devices:

```
<div class="lcd">
 <p> This text is only seen to lowest common denominator devices<p>
</div>
<div class="all">
 <p>This text is seen by all devices</p>
</div>
```

Or we can use the span element to create hooks for some devices, while hiding it from lower-end browsers. In this example, we create an empty span called "button" to use as an anchor for a complex control, like a button to toggle visibility that would be seen only on our higher-end browsers:

```
<p><a href="#show">Show all <span class="button"></span></a></p>
```

Line breaks

```
<br />, <hr />
```

Line breaks and horizontal rules work as expected on virtually all mobile devices. However, breaks should used sparingly and not as a replacement for the paragraph tag. I avoid using line breaks altogether, as they are presentation elements that should be omitted from markup, and instead use CSS to style the presentation. I use line breaks only if I know that I will use them consistently across all viewing contexts.

Character entity references

```
nonbreaking space, &, <, >, ', ", TM, ©
```

Common characters not found in the normal alphanumeric character set, such as & or ©, must be specified in XHTML as character references starting with an & and ending with a ;. Entities are often expressed in entity, decimal, or hex syntax; however, for mobile devices, the entity syntax should be used. For example, a nonbreaking space would be coded as   and an ampersand is &.

Creating Links

Links are the foundation of how hypertext works. They can take you to new pages or be used as an anchor to content further down the page. With XHTML-MP, links can also initiate a telephone call and perform other device actions in certain phones. However, due to the constrained screen size, there are additional best practices surrounding links.

Number of links

Too many links on a page makes it difficult for the user to navigate and read content. Most mobile browsers stop the vertical scroll when a link appears, meaning that for each press down on the D-pad, the user is taken only as far as the next link. Try to limit links to 10 links per page and add access keys to links whenever possible so that users can navigate with the keypad as well.

Whenever possible, try to prioritize links by popularity so that the most popular links show up at the top. This creates a better experience and ensures that important content appears above the fold. At the end of each page, the user should have someplace to go. This could be the parent category, related content, back home, or the entire navigation list—anything to help the user avoid scrolling to the top of the page for original options.

Access keys

Navigating a mobile site can be difficult and cumbersome, but you can simplify navigation and limit scrolling by providing keyboard shortcuts for common links for devices with number pads (this obviously doesn't apply to touch-only devices).

Associating an `accesskey` attribute with each link gives the user an easy way to access the link using the device's keypad. Access keys come in handy when used consistently across a site, letting users jump quickly to their chosen sections without scrolling to find a link.

You may have more than 10 links per page. Try to create access keys for all navigational links. You can save some access keys, because it isn't necessary to create access keys to links appearing within content blocks.

Initiating telephone calls

These information devices are, after all, phones, so XHTML-MP includes a means to initiate a telephone call within the `<a>` element, by prefacing the full phone number, including country code, with `tel:` within the `href` attribute. This will prompt the user to initiate a telephone call:

```
<a href="tel:+15555551212">+1 (555) 555-1212</a>
```

Because Class A browsers can render XHTML and desktop web pages, they often look for phone numbers within the page and render them as linked phone numbers, allowing you to initiate a call even if the number is not linked at all. This functionality can't always be assumed, so it is best to always link your phone numbers appropriately.

Images and Objects

The desktop web is rich with a variety of embedded content; however, due to the hardware limitation of many devices, you cannot assume that all mobile devices have the same capabilities.

Image types

Nearly all mobile devices support the JPEG, GIF, and PNG formats. Both the 8-bit PNG and the 24-bit PNG with alpha transparency are supposed to be supported as of WAP 2.0, but some older devices may not support them, due to hardware limitations. Whenever possible, use PNGs, as they are the recommended image format for the mobile web.

Image sizes

Adding images to your content can be tricky. The safe approach is to edit images so that they're as small as possible in terms of pixel dimensions. With most mobile device screens about 120 pixels wide, it is recommended that you not use images any wider than that. However, there are several devices with screens much larger.

If you are using a content adaptation system, you can dynamically insert different images based on the requesting device. Conversely, you can load larger images meant for larger devices, and use CSS to reduce the image height and width for smaller devices, but this means that lower-end devices are downloading larger images. Or you can omit images altogether from your lower-end experience, using empty spans as placeholders, and use image replacement for your higher-end devices.

Although it is a bit more complicated, I recommend the latter approach if you plan to use progressive enhancement techniques. This method allows you to control which images are loaded based on the stylesheet, even though it means defining each image manually. Lower-end devices that can't support image replacement would see no images, therefore incurring no download costs, whereas higher-end devices would see larger images sized appropriately for their display.

It might not be the perfect solution in all cases, so be creative and find the best solution for your users and their devices. Just keep in mind that every kilobyte of data you push they will likely have to pay for. Any content referenced in your markup will be downloaded, regardless of whether users see it or not. For example, if you load a 100 KB photo in your markup, but then hide it with CSS, that 100 kilobytes will be downloaded to the device, and the user will pay for data charges nonetheless. So consider the size and use of images carefully.

Image dimensions

Not specifying the pixel height and width of an image forces the mobile device to calculate the values, increasing render times and degrading performance in lower-end devices. Bitmap images have an intrinsic pixel size, so telling the browser their size in advance avoids having the browser recreate the page when it receives them. Letting the server resize the image cuts down the amount of data transferred and the amount of time it takes for the client to process and scale the image. If the specified width and height attributes match the intrinsic size, then the client doesn't resize the image.

This can, of course, be redefined using CSS, though some mobile devices might not override sizes defined in CSS correctly. If you plan to use progressive enhancement techniques, it may be easier to wait to define image sizes until you have tested them on your target devices.

Image maps

Most devices lack a pointing device, making it difficult for users to use image maps. If you know that the device supports touch, you can certainly use image maps, but I recommend avoiding them entirely.

Alt text

Always provide alt text values for all images. Downloading images can take a considerable amount of time to load over a mobile network. Some mobile browsers allow you to disable downloading images, opting for text-only mode in order to increase rendering speeds. Having alt text will ensure that any important images can be seen, regardless of the users' preferences.

Flash and SVG

Though many devices support vector objects like Adobe Flash, SVG (Scalable Vector Graphics), and SVG-Tiny, we are still a few years out from seeing these formats ubiquitously supported, due to hardware constraints. Avoid using any vector graphics unless you specifically know that the targeted devices support it.

Embedded audio and video

All WAP 2.0 devices should support the 3GPP video format and the MP4 audio format, meaning that if you link to one of these resources, it should be able to be played in the device player. However, due to hardware constraints, not all mobile web browsers support the ability to embed this content into web pages.

Tables

Before I get to how you should use tables in mobile content, I'm going to let you in on another dirty little secret of the mobile web: the best way to get a consistent layout across multiple mobile browsers has been and still is to use copious amounts of nested layout tables. Like in the days before web standard techniques on the desktop, tables provide the best way to ensure that web designs render consistency across multiple browsers.

Though the explicitness of tables means that layouts are more consistent across multiple mobile browsers, the vast array of screen widths means that sites were designed based on screen size, not device class. This approach leaves no room to create a layered, progressive enhancement approach, and limits the use to the device for which it was intended. Creating a single code source that could be rendered on each of the mobile device classes, plus gaming consoles, media centers, and the desktop, is virtually impossible.

I don't encourage you to ever employ this approach, but in the interest of full disclosure I am stating that it is still done, because it still works and in some cases it is still the best means of providing the best possible experience to multiple mobile devices. It would be easy to just adhere to web standards dogma and simply declare that there is a right way and wrong way to code your pages, but it isn't quite that easy.

Our job is to provide the best possible experience to our users, using whatever reasonable technology is at our disposal. Just because layout tables are looked down upon by the web standards community doesn't mean we should remove them from our tool chest if this just happens to be the right tool for the job.

Hopefully, you will never need to concern yourself with this legacy approach—just be aware that if you are having problems with layout consistency a table might be the solution you are looking for.

Layout tables

The web design industry considers using tables for layout as bad practice, particularly for mobile devices. Table-based layout combines presentation and markup, which makes development more difficult and essentially eliminates the ability to adapt to other media.

Though tables provide a more consistent mobile web experience, they're cumbersome and difficult to support. Table-based layouts restrict your ability to adapt for various devices and to increase page size.

It's more efficient to do page layouts with a style-based layout. The resulting layout adapts well to the narrow screens and adds flexibility while cutting page size.

Using data tables

On smaller screens, data tables—or content considered to be tabular data—often doesn't fit or can appear erratically. Unless you know that a device supports tables, avoid using them. Smaller tables with two or three columns work on most devices, but even then they're not recommended. Try using a definition list (`<dl>`) instead of a table to vertically display data.

Nested tables

Nested tables, like layout tables, don't really belong in mobile design—especially because they have a tendency to render inconsistently and add to the page size. In part because of the conditions mentioned earlier, stay away from nested tables for controlling presentation. Instead, focus on creating well-formed XHTML-MP and control the presentation with stylesheets.

Frames

Frames just don't work in mobile design. In most cases, either devices don't support them or they cause a variety of usability problems. Instead, try applying server-side includes for loading local content.

Forms

Designing and developing great forms can be a challenge in the mobile context. In the Class A browsers, forms can resemble their desktop cousins, but for all other browsers, forms don't always render like you might expect. However, the larger challenge is actually for the user, as forms are difficult to control and add content to. The rule of thumb is to limit the use of forms in the mobile context.

Free text input controls

Though unavoidable in forms that need information from the user, avoid using text boxes and text areas as much as possible. It's difficult for the user to enter content into free text input controls such as text boxes and text areas. Instead, rely on radio buttons, select boxes, and even lists of links to reduce the need for text entry.

Default input mode

It's possible to limit the type of data entered into an input field by defining the input mask or input mode using Wireless CSS or CSS-MP, thereby making it easier for users to enter information into a free text field.

The input mode (alphanumeric or numeric) of the mobile device's keypad is automatically set according to the input mask value. The following example limits the input to only numeric values:

```
<input type="text" style=' -wap-input-format: "*N"' />
```

This example limits the input to alpha characters by capitalizing the first letter:

```
<input type="text" style=' -wap-input-format: "A*a"' />
```

Other Recommendations

But wait—there's more! Here are just a few more best practices specific to mobile devices.

Validate markup

Nonvalidating markup may not display correctly or efficiently on mobile devices. In some cases, especially on older phones, nonvalidating XHTML-MP won't render at all, leaving users with an error message in their browser. Using the W3C validator can be helpful for finding rendering errors.

To check markup against the W3C mobile web best practices, you can validate your code at *http://validator.w3.org/mobile/* or *http://ready.mobi*.

Pop-up windows

Most mobile devices don't support pop-up windows. Even when they do, try not to rely on them, because changing the current view can be disorienting to the user.

External resources

Most mobile browsers download each resource as a separate element, beginning with downloading and rendering markup, followed by stylesheets and images. Depending on network speed, the user may see the basic markup while external resources download. When the download finishes, the browser renders the page again with the included elements. Carefully consider the number of external resources you use, limit them, and keep each resource's file size as small as possible.

Total page download size

Page sizes (including images and stylesheets) should remain as small as possible, because large pages take longer to load and cost the user money. Try to target your combined page weight to be between 10–25 KB when ever possible. Avoid exceeding 50 KB per page, as download times and approximate cost become increasingly prohibitive.

CSS: Cascading Style Sheets

When we are talking about inconsistencies across multiple mobile devices, what we are really talking about is CSS. In the past, mobile devices had incredibly poor support for CSS, using it as nothing more than a way to style text and apply background colors. Though many of today's mobile browsers have far better support for both CSS2 and CSS3 than their predecessors, there are still plenty of legacy mobile browsers in the market to contend with.

Designing your CSS to work across multiple mobile browsers isn't easy and can be quite a painful process. There is no one perfect way to create CSS that renders consistently on more than a handful of devices. I have three techniques I use, depending on the devices I intend to support:

Keep it simple
> Keeping your styles very basic, using no complex styling techniques whatsoever, can be the ideal method for simple sites. Though it may not be pretty to look at, it works.

Code and reload
> In this approach, you constantly test how your styles render on devices. For each code change, you reload the browser on each device you plan to support. This approach is slow and tedious, but it means fewer issues toward the end of the project.

Progressive enhancement
> As I discussed before, this approach requires you to create multiple layers of support, so that your style gracefully degrades depending on the device. This technique takes some practice to get right, but if you can master it, it can be a powerful approach.

Wireless CSS and CSS-MP

For markup, we have XHTML-MP, a descendant of XHTML; it only makes sense that we have a mobile equivalent for CSS. In fact, we have two: Wireless CSS (sometimes referred to as W-CSS or WAP CSS) managed by the OMA and part of WAP 2.0 along with XHTML-MP. And then we have CSS-MP, or CSS Mobile Profile, managed by the W3C. So, in case you are keeping score: the OMA owns XHTML-MP and W-CSS, and the W3C owns XHTML Basic and CSS-MP.

The good news is that both of these standards are working to come together into one standard; they are both based on CSS2.1, and they are both becoming somewhat irrelevant, as mobile browsers gain decent support for CSS2.1, not to mention CSS3.

The primary additions are a handful of properties meant to improve the user experience, including a few input masks and marquee controls. If you can live without these, there is really no need to get too concerned with these mobile-only standards.

Box Model

The box model is one of the key concepts of CSS design, and therefore the first thing that tends to go wrong in mobile devices. The box model is the imaginary box that is around every element in your markup. It consists of five areas: the content, the padding, the border, the margin, and the outer edge, as shown in Figure 11-3.

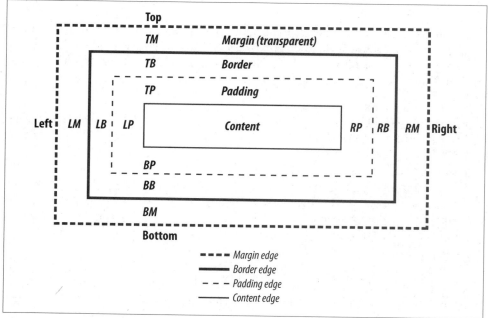

Figure 11-3. The box model

Many elements have inherited values, meaning that they may have some margin or padding by default. The paragraph tag for example has inherited margins above and below the content area. Due to the space constraints, mobile browser engines sometimes apply different inherited values than you might expect. This can lead to frustrating inconsistencies among multiple mobile browsers.

In Table 11-2, you can see how the box model compares with each class of mobile browser.

Table 11-2. Box model comparison

	Class A	Class B	Class C	Class D	Class F
Box model	Great	Good	OK	Poor	Fail

Selectors

The selector is used to tell which markup elements it should apply rules to—basically, what makes CSS work to control the presentation. There are a number of different types of selectors:

- Universal
- Type or element selectors
- Contextual selectors (descendant, child, and adjacent sibling)
- Class and ID selectors
- Pseudoclasses
- Pseudoelements

Universal selector

The universal selector selects all elements useful for defining the default typeface or font size. body or html can be used as well:

```
{font-family:serif;}
```

Type selector

The type or element selector targets the element by name. Type selectors are fairly safe to use across all mobile devices:

```
h1 {color: red;}
```

Descendant selector

The descendant selector targets elements that are descendants of another element, indicated by a single space separation. Despite being a fairly standard technique for styling desktop sites, the descendant selector is not consistently supported on Class C or lower browsers:

```
li a {color: black;}
```

Child selector

The child selector is similar to the descendant selector, but targets only the child of the defined element, not all instances like the descendant selector. Like the descendant selector, Class C browsers and lower can render child selectors inconsistently:

```
p > em {background-color: yellow;}
```

Adjacent sibling selector

The adjacent sibling selector targets an element that comes immediately after another element with the same parent. Again, this not a widely supported selector for Class C or lower browsers, but you can find some incompatibility among Class B browsers as well:

```
h1 + p { margin-bottom: 1em;}
```

Class selector

The class selector targets any element with the matching class. Class selectors are one of the more common techniques used for cross-platform mobile designs, given their wide support in nearly all mobile browsers:

```
.error {background-color: red;}
```

ID selector

The ID selector targets any element with the matching ID (remember that all IDs need to start with at least one letter). ID selectors, like class selectors, are one of the more common techniques used for cross-platform mobile designs, thanks to their wide support in nearly all mobile browsers:

```
#alert {background-color: yellow;}
```

Simple attribute selector

A simple attribute selector targets all elements with the matching value. These attribute selectors tend to work only in newer mobile browsers:

```
p[class] {color: blue}
```

Advanced attribute selector

Advanced attribute selectors target elements with matching substrings in the value of the attribute. This means that you can target existing attributes or create your own custom attributes. Although not supported on many Class B or lower browsers, advanced attribute selectors can be quite helpful for targeting different device classes, using a custom attribute:

```
<p device="iphone"> iPhone </p>
<p device="android"> Android </p>
<p device="lcd"> Lowest Common Denominator </p>
```

In this example, the caret (^) is used to target the attribute that begins with the value "iphone":

```
p[device^="iphone"] {background: green;}
```

In this example, the dollar sign or string ($) is used to target the attribute that ends with the value "android":

```
p[device$="android"] {background: blue;}
```

In this example, the star (*) used to target the attribute that contains at least one instance of "lcd":

```
p[device*="lcd"] {background: red;}
```

Pseudoselectors

You can use pseudoselectors to target elements that might not exist in the markup, like a visited link. Unfortunately, pseudoselectors do not have strong support in Class B or lower devices:

```
:link, :visited, :hover, :active, :before, :after, :first-child, :last-child
```

In Table 11-3, you can see how different CSS selectors fare in each browser class.

Table 11-3. CSS selector compatibility table

Selector type	Class A	Class B	Class C	Class D	Class F
Universal selector	Yes	Yes	Yes	Yes	Yes
Type selector	Yes	Yes	Yes	Yes	Flaky
Descendant selector	Yes	Yes	Flaky	Flaky	No
Child selector	Yes	Yes	Flaky	Flaky	No
Adjacent sibling selector	Yes	Yes	Flaky	No	No
Class selector	Yes	Yes	Yes	Yes	Flaky
ID selector	Yes	Yes	Yes	Yes	Flaky
Simple attribute selector	Yes	Flaky	No	No	No
Advanced attribute selector	Yes	No	No	No	No
Pseudoselector	Yes	Flaky	No	No	No

Font and Text Properties

The typography options on mobile devices can be less than stellar, but like most things CSS-related, we are seeing mobile browsers move closer to their desktop cousins in this respect. This section covers the font and text options for mobile devices.

Available fonts

With the desktop web, we have at least 10 different fonts we can use in our designs. In mobile development, we can count on only two options: serif and sans-serif. In low-end devices, we might have only one option—almost always a sans-serif variant:

```
p {font: sans-serif;}
```

Traditionally, in mobile devices the only font available was the device font, and there was really no point trying to define a font at all. Over the years, the typography choices have become a bit more diverse, adding web-safe fonts like Georgia, Times New Roman, Arial, and Helvetica, and a monospace font like Courier New, but we see these options now only in Class B and higher browsers.

Absolute size keywords

It's not very common in desktop web design, but the typical method to determine the size of text is using one of seven absolute size keywords: xx-small, x-small, small, medium, large, x-large, and xx-large:

```
p {font-size: xx-small;}
```

The bit depth of mobile screens can vary widely, so keywords are useful; they do not correspond to a precise measurement. Instead, they are relative to one another. Therefore, while one mobile browser may have a default text of 11 pixels and another 14 pixels, by using absolute size keywords, the medium keyword would be the default and recommended text size for both devices.

Percentage measurements

Percentage size values will work on Class B and higher browsers, but I wouldn't recommend them:

```
p {font-size: 80%;}
```

Percentage size values rely on inherited parent values, which are not consistent across all browsers. Use absolute size keywords instead.

Length measurements

One of the more common means to style text on the desktop web is by using a specific number of units—for example the pixel, which is relative to the display resolution or ems, or the distance from baseline to baseline:

```
p {font-size: 10px;}
p {font-size: .9em;}
```

Though possible on Class B and higher mobile browsers, it isn't recommended. Again, the screen's pixel depth can vary from device to device, creating inconsistent, unreadable designs.

Additional styling

There are a great number of ways to style text. The majority of them will work across multiple devices, given that if there is one thing a mobile browser should do fairly well, it is display text:

```
font-weight: bold;
font-style: italic;
text-transform: capitalize;
text-decoration: underline;
line-height: 2em;
text-align: right;
white-space: nowrap;
```

The general rule of thumb is that if you keep it simple, it will work across the majority of devices, but you can have peace of mind knowing that most basic text styling techniques will work as expected.

Text shadow

Creation of a text shadow, though supported only by Class A browsers, is a useful technique for mobile devices, reducing the need for images to create a desired visual effect:

```
text-shadow: 2px -1px 2px #ccc;
```

The syntax is: *x-coordinate* (2px) *y-coordinate* (−1px) *blur-radius* (2px) *color* (#ccc).

Font replacement

Given the limited typography options available in the desktop web, many designers and developers are starting to use text replacement techniques. There are three techniques used—some that work on Class A mobile browsers only:

@font-face
> This easiest method is to define a font using the CSS3 at-rule @font-face, which references the typeface file to be used. But due to the heavy processing required to render custom fonts, this technique is not yet widely supported.

sIFR
> The technique uses Flash to replace the text with a small vector representation of the desired glyphs. However, this technique requires a browser with the Adobe Flash Lite plugin.

canvas
> The third technique is to use JavaScript and the canvas element to render text to the device. I've found this to be the only reliable way to replace text on mobile devices, but only for browsers that have good JavaScript support and that support the canvas element.

In Table 11-4, you can see the different font support for each class of browser.

Table 11-4. Font and text compatibility

	Class A	Class B	Class C	Class D	Class F
Available fonts	Web-safe fonts	Web-safe fonts	Sans-serif and serif	Sans-serif and serif	Sans-serif and serif
Font size	Any	Any	Keyword	Keyword	Keyword
Font-weight	Yes	Yes	Yes	Yes	Limited
Font-style	Yes	Yes	Yes	Yes	Limited
Text-transform	Yes	Yes	Yes	Yes	Limited
Text-decoration	Yes	Yes	Yes	Yes	Flaky
Line-height	Yes	Yes	Yes	Yes	Flaky
Text-align	Yes	Yes	Yes	Yes	Yes
White-space	Yes	Yes	Yes	Limited	Flaky
Text shadow	Yes	No	No	No	No
Font replacement	Limited	No	No	No	No

Basic Box Properties

Being able to style the box area around an element is a crucial part of web standards design. The good news is that the basic CSS level 2 box styling techniques you might use for the desktop web do work on most mobile devices, allowing you to style content with some level of precision. Many of the techniques that are not fully supported, like percentage-based and min-height techniques, are the same ones that many desktop web browsers don't fully support.

Height and width

Height and width can be specified on the majority of mobile browsers, but issues can occur when trying to define percentage-based height values:

```
width: 100%;
height: 50px;
```

Minimum and maximum dimensions

Minimum and maximum dimension values are not a reliable means to style an element across multiple mobile browsers. Though it is a very useful technique in the mobile context, it is largely used for Class A browsers only:

```
min-width: 120px;
max-height: 100px;
```

Margins

The margin is the area applied to the outside of the element, including the border area. Luckily, margins render fairly consistently and can be relied on across multiple mobile browsers:

```
margin: 10px;
```

Padding

Padding is applied to the area within an element, inside of the border area. Like margins, padding works across multiple mobile browsers, however, not as reliably in lower-end browsers that have flawed box models. However, this can be worked around by applying margins to a child element:

```
padding: 10px;
```

Borders

Simple borders, like this one, will work in many mobile browsers, but can sometimes cause some inconsistencies in low-end browsers that have flawed box models:

```
border: 1px, solid, black;
```

Some Class A browsers do support advanced CSS3 border techniques like rounded corners, border images, and box shadow.

Box shadow

Like text shadowing, the box shadow is a useful way to create a desired visual effect without relying on downloading images to the device, though it is supported only by Class A browsers:

```
box-shadow: 10px 10px 5px black;
```

The syntax is similar to that of the text shadow: *x-coordinate* (10px) *y-coordinate* (10px) *blur-radius* (5px) *color* (black).

In Table 11-5, you can see how each of the box properties stack up against each of our browser classes.

Table 11-5. Box properties compatibility

	Class A	Class B	Class C	Class D	Class F
Height and width	Yes	Yes	Limited	Flaky	Flaky
Min and max dimensions	Yes	Flaky	No	No	No
Margins	Yes	Yes	Yes	Yes	Limited
Padding	Yes	Yes	Yes	Flaky	Flaky
Borders	Advanced	Limited	Limited	Flaky	Flaky
Box shadow	Yes	No	No	No	No

Color and Backgrounds

Styling an element means defining colors and background images. Relying on CSS instead of images to create desired visual effects reduces time to download as well as cost.

Background color

The background color allows you to add a color value to the content area of an element. This technique works fairly reliably across all mobile devices:

```
background-color: red;
```

Background image

The background image allows you to link an image to the content area of an element:

```
background-image: url(background.png) repeat-x;
```

The `background-image` property is incredibly useful for creating mobile designs, allowing you to send small images to the device and tile them in a number of ways to create bandwidth-friendly designs.

Multiple background images

Though supported by Class A browsers only, the multiple background support of CSS3 comes in quite handy when creating mobile designs. It is written exactly the same as the `background-image` property; you can add multiple source values separated by commas, the first background being the foremost and the last being the furthest back:

```
background-image: url(foreground.png) no-repeat, url(background.png) repeat-x;
```

Table 11-6 displays how each of the color and background attributes are supported in each class of browser.

Table 11-6. Color and background compatibility

	Class A	Class B	Class C	Class D	Class F
Background color	Yes	Yes	Yes	Yes	Yes
Background image	Yes	Yes	Yes	Flaky	Flaky
Multiple background images	Yes	No	No	No	No

Positioning and Page Flow

CSS goes beyond just being able to add design content within the page; it can also be used to define the design layout of the page. Using positioning and page flow attributes, we can add style to the page and help make it easier to read or interact with on small screens.

Display

The `display` property allows you to define how an element is to appear in the page flow.

```
display: block;
```

For example, if the value is `display:none`, then the element is removed from the page flow and hidden from the user. This technique is often used in the mobile context to hide elements from the page flow that do not apply to the viewing context, for example in the case of hiding extraneous content meant for lower-end browsers.

The toggle display property with JavaScript is one of the principle tests for browser classification, allowing the designer to create dynamic tabs, or lists that the user can interact with to reveal. This technique allows the designer to conserve space on-screen and provide clearer direction to the user.

Floats

Floats are usually applied to images and other elements to position them either to the right or left of the page flow, causing nearby elements to wrap around them. Due to the smaller screens of mobile devices, floats are not encouraged for layout use. Instead, try to limit the use of floats to inline elements you wish to display alongside related text:

```
float: left;
```

Clearing

Clearing goes hand in hand with floats, allowing you to prevent how block-level elements wrap in the use of a float. Usually, support for clearing in mobile devices is added to the browser whenever it adds support for floats:

```
clear: left;
```

Positioning

Both relative and absolute positioning might seem to be odd tools for mobile design given the small screen size, but they actually come in handy for creating designs for Class B and higher browsers:

```
position: absolute;
top: 10px;
left: 50px;
```

In this example, absolute positioning removes the element from the page flow and places it at the specific position relative to its container.

Due to the way different mobile browsers treat the viewport, fixed positioning can be somewhat unreliable, even in Class A browsers.

Overflow

There are several types of overflow: visible, hidden, scroll, and auto. This property is designed to control how content is displayed when it exceeds the defined value of an element:

```
overflow: hidden;
```

Both the visible and hidden types are fairly well supported in Class B and higher browsers, allowing you to save space in lists and menus. However, scroll and auto (which defaults to scroll in the case of an overflow) are not treated consistently across many browsers. The reason is that it is difficult to provide the user with an established means of scrolling the area on a device.

Stacking order

The z-index element is used to determine the stacking order of positioned elements:

```
z-index: 10;
```

This property is used when overlaying elements on top of each other to create a specific visual effect. Though used mostly for designs targeting Class A browsers, it does work on other classes as well.

Table 11-7 shows how positioning and page flow attributes are supported in each of our mobile browser classes.

Table 11-7. Positioning and page flow compatibility

	Class A	Class B	Class C	Class D	Class F
Display	Yes	Yes	Yes	Yes	Flaky
Toggle display	Yes	Yes	Limited	No	No
Floats	Yes	Yes	Limited	Limited	Flaky
Clearing	Yes	Yes	Limited	Limited	Flaky
Positioning	Yes	Yes	Limited	Flaky	No
Overflow	Yes	Limited	Flaky	No	No
Stacking order	Yes	Yes	Limited	Flaky	No

JavaScript

Last, but not least, we come to JavaScript: the last pillar of mobile web development. Unfortunately, JavaScript simply hasn't been a priority in mobile browsers for many years, due to the hardware limitations of devices. I frankly consider myself lucky to be dealing with a device that has decent CSS support, and if I have devices that support a little bit of JavaScript as well, then I'm ecstatic.

The bad news is that unless you are targeting Class A browsers, JavaScript just can't be assumed in your mobile project. I know this can be a potential barrier to entry for many mobile web projects, but it doesn't have to be. Just go back to the concepts of progressive enhancement that I talked about at the beginning of this chapter, and start to map out how your experience will degrade to each of the device classes you plan to support.

The good news is that we are starting to see widespread JavaScript support in multiple mobile browsers with Class A browsers and a few Class B browsers. Some devices, like the iPhone, are taking JavaScript to a new art form (which I will discuss more in the next chapter). As devices get more powerful and browser engines more adept at processing JavaScript, I have a feeling that this problem will go away soon.

Table 11-8 shows that JavaScript is available only in a few devices.

Table 11-8. JavaScript compatibility

	Class A	Class B	Class C	Class D	Class F
JavaScript support	Yes	Some	No	No	No
DHTML	Yes	Limited	No	No	No
Ajax	Yes	Limited	No	No	No

iPhone Web Apps

I've said it many times already: the iPhone was a game changer in the mobile ecosystem. A big part of that change is the impact that it had on the mobile web and specifically on mobile web applications. The iPhone provided people all over the world a glimpse of the future of mobile—that the mobile web didn't have to be these ugly lists of text, that now we could create something unique and cool for mobile devices using the same techniques we use on the Web.

I have to admit that I've deceived you with the title of this chapter; this chapter isn't about creating iPhone web apps, it is about creating mobile web apps for the iPhone and beyond. Because when we are talking about iPhone web apps, we are actually talking about WebKit, the mobile web browser behind the iPhone and iPod touch, and also the device browser in some of the best-selling smartphone platforms, like the Nokia Series60 (or S60), Android, Palm's webOS, and more. In a short period, WebKit went from being just the core technology for Apple's web browser Safari to one of the top, most proven mobile browsers in the world.

But the story doesn't stop with just WebKit. The iPhone created a sea change within the entire mobile web landscape. After the iPhone, the device browser suddenly went from being a third-class citizen to being its killer app. Operators and device makers partnered with browser makers to get competitive browsers in their devices to rival the features of the iPhone. Once the iPhone bolted out as the market leader, the rest of the pack knew exactly what they needed to do to catch up, or in some cases compete: what the market wanted.

The secret to WebKit's mobile successes isn't the fact that it powers the device browser of the iPhone; it is simply that it has excellent support for defined web standards. Support for web standards, and the ability to bring a desktop-quality web browser to mobile devices is out of reach for all but a handful of mobile browsers. Ironically, we've learned since the iPhone's debut that excellent support for web standards was the basic obstacle that had held back massive adoption of the mobile web. WebKit brought the Web that we've come to expect and love from the desktop to our mobile devices. In doing so, it has defined a new category of mobile content: the mobile web app, which barely existed before the iPhone. This chapter dives into what makes iPhone web apps tick. It looks at WebKit and introduces some techniques that will work only on WebKit and some that will work only on an iPhone or iPod touch. This isn't to play favorites to Apple's devices, but to showcase how the market is being defined and applied to devices beyond the iPhone.

Why WebKit?

A common question during my talks about mobile design and development is why I am such a big proponent of the iPhone and WebKit. If there are a number of other browsers that are actually technically superior, support even more standards, or even have a higher market penetration, how can I in good conscience recommend prioritizing development around the iPhone first, and in some cases recommend solely focusing on the iPhone?

Here are a few reasons why I believe the iPhone is worth prioritizing:

- The iPhone has proven to be the market leader in terms of innovation and market share of the devices that access the mobile web the most.
- The iPhone is marketed, sold, and supported by Apple, not the operator— operators tend to promote phones for only short periods of time, meaning that devices need to become popular in their own right, or the operator needs to push them. The iPhone is marketed, sold, and supported by Apple in addition to the operator. Other device makers attempt to do this, but Apple's strong reputation and brand appeal make it unique.
- It has the lowest development cost for the highest number of supported devices— very little is needed to create a successful iPhone web app. It takes only a text editor and a web browser. No devices, emulators, or special software are required.
- It requires little to no additional knowledge apart from HTML, CSS, and JavaScript.
- It has a simple and cost-effective testing environment—device testing can be done on an inexpensive iPod touch on a Wi-Fi network and render exactly the same on an iPhone.
- The highest consumer of mobile data—every trend we are seeing shows that users that have devices that run WebKit use the mobile web a lot more often and for longer periods than your typical mobile user.

A Brief History of WebKit

WebKit is an open source web browser engine derived by Apple from the Konqueror HTML layout engine called KHTML and its KJS JavaScript engine. Apple ported KHTML and KJS to Mac OS X to create the first Safari desktop browser in 2003 and has since been used in a number of Mac OS X applications.

You might not know him by name, but one of the key contributors and architects of WebKit is Dave Hyatt, whom I consider to be one of the great influences on the modern Web today. While working at Netscape, Hyatt created a Mac OS X application called Chimera that was based on Mozilla's Gecko rendering engine. At the time, Netscape/Mozilla was a hefty and bloated application, rapidly losing what market share it had left to Microsoft's Internet Explorer. Chimera, later to be renamed Camino, was at the time the fastest web browser on any Mac or PC, showing that the Gecko could be turned into a fast, lightweight browser. Camino also introduced or popularized many features commonplace in the web browser, like tabbed browsing and pop-up blocking.

The PC world was a bit jealous with the power and speed of Camino, and thus the aptly named Phoenix—later to be labeled Firefox, and cocreated by Hyatt—was born, going on to become the second most popular desktop browser today.

Firefox, thanks to Hyatt, inspired a new generation of web designers and developers to shift how they went about creating websites. Without this shift, we would not have seen the popularization of Ajax, which made web applications possible and spawned the entire Web 2.0 movement. In 2002, Apple hired Hyatt to work on the WebKit project.

Though it may seem like a minor historical footnote, I view it as crucial evidence of the importance of Dave Hyatt and his employer, Apple, to the future of the browser landscape. Together, they have changed the way we interact with the Web on the desktop and on mobile devices.

Background As a Mobile Browser

WebKit's life as a mobile browser engine did not start at Apple; it started at Nokia, the number-one device maker in the world, long before anyone outside of Apple knew about the iPhone. WebKit is ideal for mobile devices, given its small footprint and capability to render full-scale desktop web designs to mobile devices. And with titans like Apple, Google, and Nokia all returning innovations to the open source project, WebKit means that device makers or enterprising young browser makers looking to make their mark on the mobile web landscape can dive right in with a powerful and tested mobile browser without having to start from scratch.

The following sections discuss the known mobile browsers powered by WebKit.

Web Browser for S60

The Web Browser for S60, or the Nokia Mini Map Browser (Figure 12-1), is the default mobile browser for the S60 platform, common in most N and E series Nokia smartphones. The browser uses a Mini Map, for which it is nicknamed, to allow the user to pan around and zoom into a web page not optimized for mobile devices.

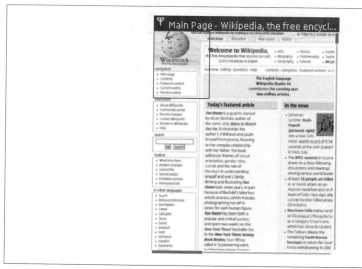

Figure 12-1. The Nokia Web Browser for S60, which is based on WebKit

iPhone and iPod touch

WebKit made its big splash in the mobile market as the device browser for the iPhone and iPod touch, usually referred to as Safari or Mobile Safari (Figure 12-2). Apple's version of WebKit includes the ability for users to double-tap on an area of a nonoptimized web page; it then zooms in and out of the area. Users can pan around the page with their finger, rotate the orientation, or use the multitouch gestural features, like pinching to zoom in and out.

Because the iPhone initially supported only web apps, not native applications, many applications were built largely using WebKit-specific styles meant to work exclusively in Mobile Safari.

Android

Included in the open source mobile framework Android is a WebKit-based device browser. Because Android is open source and can be customized by the device maker or the operator, the WebKit browser can be replaced by another browser written for the Android platform, though no one has announced plans to do so yet. Many of the top device makers have released or plan to release devices based on the Android platform.

In addition, within the Android Market, the app store for Android-compatible apps, there are a number of additional browsers built on WebKit available for free download—some offering additional features more akin to those on Mobile Safari, as well as newer versions of the WebKit trunk than what ships in Android (Figure 12-3).

Figure 12-2. Safari on the iPhone, sometimes referred to as Mobile Safari, which is based on WebKit

webOS

Palm has created an entire operating system based on WebKit, called webOS. Using the Mojo SDK, developers can create "native" applications using HTML, CSS, and JavaScript. All applications render using the WebKit browser engine (Figure 12-4).

Myriad Browser

The Myriad Browser v9 is a browser from the Myriad Group that is also based on WebKit. Myriad purchased the Openwave Mobile Browser from Openwave, which over the years has released some of the most popular mobile web browsers in the world. Myriad, like most mobile browser makers, works with device makers and operators to preload devices with their product, which is often referred to as the device browser.

Iris Browser

The Iris Browser from Torch Mobile is one of the many WebKit browsers designed for Windows Mobile devices. Torch Mobile works with operators to preload the Iris Browser on Windows Mobile devices, and is not meant to be a direct-to-consumer

Figure 12-3. The Android browser, which is based on WebKit

product, though Windows users can download a free copy from the Torch Mobile website.

What Makes It a Mobile Web App?

Technically, any content viewed in a web browser can be considered a website, but we are certainly seeing a subset of experiences that are defined in the mobile context as web applications, or web apps. Apple has taken this to heart with Safari on the iPhone and WebKit, adding several proprietary tricks to tell Mobile Safari that the content in view is a web app intended for the iPhone, and not just a website.

The popularity of mobile web apps for the iPhone and iPod touch has caused many other mobile browser makers to emulate Apple's techniques, but some are introducing their own proprietary methods. The W3C is working on defining best practices for mobile web applications, but it could be a while until they are agreed on and adopted by the mobile browser makers. Until then, my advice is to look at how Apple defines

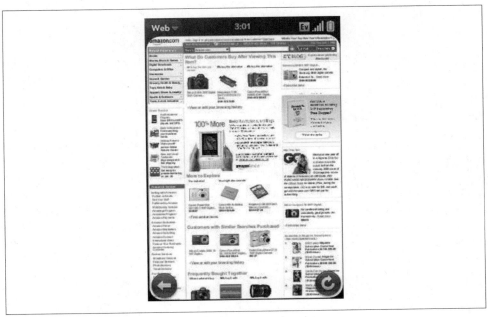

Figure 12-4. The Palm Prē running webOS, based on WebKit

web apps from websites, as they will be the market leader in this area at least for the foreseeable future.

The following are some common characteristics of a mobile web app:

- Is an application-like experience that alters existing views, in place, instead of loading new pages like a traditional website
- Uses client-side (or offline) storage for local data
- Heavily uses DHTML or Ajax to create the user experience
- Has a defined viewport for the mobile context
- Can run in full-screen mode
- Can be launched like a native application

The Page Model

The key difference between a mobile website and a mobile web app is the type of page model that you use. Mobile web apps usually use a single-page model in which the browser loads one container page of markup. Then, using Ajax, additional content is loaded based on the user's actions. The user can perform most if not all tasks from this one coded page. In the multipage model, the user traverses a hierarchy of individual pages, designed to lead the user to the desired end goal. In this model, the user clicks back and forward controls to recall content, whereas in the single-page model the browser's back button is not meant to be used.

Both of these models can be used together to create a hybrid app, which most would still consider to be a web app; however, due to how the iPhone can render content that is explicitly defined as a web app, which I discuss in the following pages, I recommend selecting only one model. If you select the single-page model, the problem becomes that your content no longer gracefully degrades to lower-end devices. Effectively, if the device browser doesn't support Ajax, your content won't be viewable.

It is possible to get around this limitation using server-side logic to render page headers dynamically depending on the requesting device, enabling you to render Ajax-friendly pages to iPhones and other devices with Class A browsers, while rendering a full document for lower-end browsers.

For the rest of this chapter, I will discuss how to leverage the iPhone, iPod touch, and other WebKit-based browsers to create applications using XHTML markup, CSS, and JavaScript.

Markup

WebKit is about as modern a browser as it gets, supporting nearly every approved web standard we have; the only other browser engine that comes close is the Trident browsing engine used in many of Opera's browsers. The challenge with today's browsers is for the first time, desktop browser makers are ahead of the W3C, implementing standards that are being discussed by the W3C working groups but not yet finalized. My hope would be that by the time you read this book, the specifications for HTML5 or XHTML2 (as well as CSS3) will have been finalized, but I'm not going to hold my breath. The debate and discussion of these new standards has been going on for ages, and innovation simply can't wait for the final specification to be handed down. This means that browser makers are moving forward implementing what and how they think is the best way to go about it, then reporting their findings back to the working groups—WebKit included.

XHTML

When the HTML standard was published in 1990, it was designed for formatting and laying out elements on the screen contextually. For the most part, it was logical, but there were some exceptions that made parsing it quirky. For instance, the open line-break tag <br> did not need to be closed.

A few years later, the similar XML syntax standard for marking up data semantically came out. Because the structure and content of the data is unknown, XML is much stricter about its structure.

Though the two formats look syntactically similar, they have different purposes. XHTML is HTML that comes with the benefits of conforming to the stricter syntax of XML. Further, it combines the contextual markup of HTML with the semantically structured data of XML.

Over the years, the exceptions in HTML have slowly given way to correction, and now properly written HTML can often be validated against the XHTML standards. About all that is needed is to change the DOCTYPE specifier before the root element.

The iPhone and WebKit support the XHTML 1.0 Strict and Transitional doctype, which is the recommended language for writing iPhone web apps. Virtually all XHTML elements and techniques are supported on the iPhone and WebKit-powered browsers—yes, even the blink tag.

With desktop sites, the trend for the past several years has been to write well-formed, semantically coded XHTML, with a minimum of presentation-related code, meaning no layout tables or single-pixel GIFs meant to control the layout. By having the presentation removed from our markup, we can display our content in a variety of different ways—on a desktop browser, a mobile browser, a gaming console, and so on, each designed for the context it is intended for, with a minimum of rewrites. This is known as *web standards*, but actually refers to principles and techniques, not to how well your code validates according to the W3C specifications.

Here is a simple example of what "web-standard code" might look like:

```
<!DOCTYPE html PUBLIC "-//W3C//DTD XHTML 1.0 Transitional//EN"
"http://www.w3.org/TR/xhtml1/DTD/xhtml1-transitional.dtd">
<html xmlns="http://www.w3.org/1999/xhtml" xml:lang="en" lang="en" dir="ltr">
<head>
  <title>An XHTML 1.0 Compliant Document</title>
</head>
  <body>
    <p>Here is some text</p>
  </body>
</html>
```

Traditionally, in mobile writing, standard markup just didn't work, or at least in the most popular devices sold. What made the iPhone unique is that it changed that perception among web designers and developers, in that they could use the exact same principles and techniques they used for desktop sites and for mobile sites and apps. Therefore, if you know XHTML, you can transition your knowledge to creating iPhone web apps, instantly. There is virtually no learning curve for writing simple iPhone web apps in XHTML.

In fact, Apple recommends that the developer allow the user to browse the desktop version of the site. If an iPhone version of the site is available, you should provide a link to it rather than automatically redirecting the user there. This makes sense for future-proofing your site for new versions of WebKit-based devices.

Assuming that you provide the user with well-structured XHTML 1.0–compliant documents, you can detect the iPhone on either the server or the client. The implementation is up to you. This book implements a client-side detector, because it works regardless of the backend.

A lot of web sources incorrectly suggest sniffing the client using JavaScript and then testing. Though it's probably easier to implement, this technique broke when the iPod touch was released. Instead, the solution recommended by Apple is:

```
function iPhoneClient()
{
  return RegExp(" AppleWebKit/").test(navigator.userAgent) &&
    RegExp(" Mobile/").test(navigator.userAgent);
}
```

This snippet tests the browser's identification string (the **userAgent**) for two strings: "AppleWebKit" and "Mobile". Elsewhere, this function can be used to, say, swap out the CSS stylesheet for an iPhone-specific version, or, as Apple suggests, display a link to an iPhone-specific version of the site.

On the upside, this sniffer doesn't look for the "iphone" string specifically, so it will work for the iPhone, iPod touch, and any future devices based on WebKit.

However, this is a double-edged sword—there are some downsides to this technique. First, although this snippet will return "true" for the iPhone and the iPod touch, it will also return "true" for other WebKit mobile browsers. Some non-iPhone devices will be erroneously detected as iPhones. Secondly, though the iPhone at the time of this writing supports only the built-in WebKit Safari for iPhone, at some point Apple may allow installation of browsers based on other platforms, such as Gecko. This sniffer would not detect those browsers.

XHTML-MP

The iPhone will render XHTML Basic and XHTML-MP pages, but it won't like it (Figure 12-5). Given the option to render a desktop version, or a mobile or WAP version of a site, the iPhone will render the desktop version. Apple considers the iPhone to be a "One Web" device, meaning that it will always try to choose the best possible experience over the lowest common denominator, opting instead to use media queries to define alternative stylesheets designed for the iPhone. Most WebKit-based mobile browsers, with the exception of the Nokia Mini Map browser, follow this principle.

XHTML 2.0

At this time, WebKit does not support XHTML 2.0.

XHTML 2.0 is the successor to XHTML 1.0, the primary markup language of web standards these days, including most iPhone web apps. XHTML 2.0 is based on the principles of XML and is meant to be a minimal, all-purpose language used for marking up content to be machine-readable, regardless of context. Using CSS for the

Figure 12-5. Testing XHTML-MP code in WebKit

presentation, or using the content strictly as a data source, like what we can do with an RSS or ATOM feed, we could adapt designs to multiple contexts.

This all sounds great in theory, but XHTML 2.0 would be a massive change to how we create sites and web applications. Instead, most browsers—WebKit included—seem to be moving more toward HTML5. There are many similarities between HTML5 and XHTML 2.0, and I expect that we will see HTML5 as more of the transitional standard while XHTML 2.0 remains the pie-in-the-sky goal.

HTML5

HTML5 is designed to be the successor to HTML4, as a transitional standard to XHTML 2.0. It stays fairly close in syntax to the HTML4 and XHTML 1.0 standards that we are accustomed to, but moves away from its SGML roots as being simply a means to mark up a document, and into a robust web application platform, including several new scripting APIs that allow for common interactive functions in today's web applications.

Originally referred to as Web Applications 1.0, HTML5 incorporates the Web Forms 2.0 standard from the W3C's Web Hypertext Application Technology Working Group (WHATWG). The editors of the specification are Ian Hickson of Google and Dave Hyatt from Apple—and it's probably not a coincidence that these are the same two companies that produce both mobile and desktop versions of WebKit.

HTML5 creates some interesting new opportunities for mobile web applications, like the `canvas` element, offline storage, document editing, and media playback, which we are already beginning to see in mobile WebKit browsers like Mobile Safari as well as the Opera Mobile browser. HTML5 also allows developers to create cross-platform designs through expressing the content more semantically. For example, the addition of HTML5 elements like `header`, `nav`, `article`, `section`, `aside`, and `footer` make our content more machine-readable and therefore make it easier for the next generation of mobile browsers to treat content properly in both the desktop and mobile context, as shown in Figure 12-6.

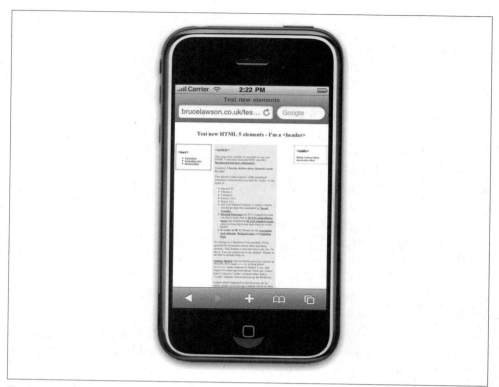

Figure 12-6. Testing HTML5 code in WebKit

Because these standards are still being defined and the device browsers are being updated to support them, check the Wikipedia page, which is actually quite up-to-date (*http://en.wikipedia.org/wiki/Comparison_of_layout_engines_(HTML_5)*, to check the status from the desired device to see the latest HTML5 APIs and elements that are currently supported.

The canvas element

The canvas element is part of HTML5; it allows designers and developers to essentially draw content within your HTML page. The canvas HTML tag defines a custom drawing area within your content that you can then access as a JavaScript object and draw upon. canvas was created by Apple, included in the WebKit source and the iPhone, but it is also supported by the Mozilla Gecko and Opera Presto browser engines.

Here we draw some simple boxes using the canvas element, as shown in Figure 12-7:

```
<head>
  <script type="text/javascript" charset="utf-8">
  function draw()
  {
    // grab the canvas element
    var canvas = document.getElementById("canvas");
    if (canvas.getContext)
    {
      // grab the context
      var ctx = canvas.getContext("2d");
      // background box
      ctx.fillStyle = "rgba(100, 100, 100,0.2)";
      ctx.fillRect(0, 0, 90, 90);
      // first, smallest
      ctx.fillStyle = "rgba(100,100,100,0.5)";
      ctx.fillRect(10, 10, 10, 10);
      // second, middle
      ctx.fillStyle = "rgba(100, 100, 100,0.7)";
      ctx.fillRect(20, 20, 20, 20);
      // third, biggest
      ctx.fillStyle = "rgba(100, 100, 100,0.9)";
      ctx.fillRect(40, 40, 40, 40);
    }
  }
  </script>
</head>

<body onload="draw();">

  <canvas id="canvas" width="150" height="150">
    <p>This example requires a browser that supports the <a
href="http://www.w3.org/html/wg/html5/">HTML5</a> &lt;canvas&gt; feature.</p>
  </canvas>

</body>
```

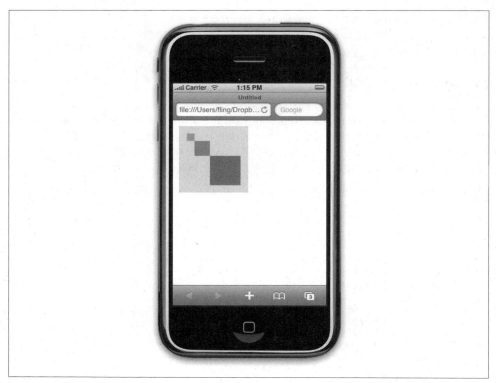

Figure 12-7. Using the canvas element on the iPhone

Offline data storage

Also part of HTML5 and supported by WebKit and the iPhone is the ability to create client-side data storage systems, which essentially allow you to create web applications that work when offline. Because it is still evolving into a standard, client-side storage has a variety of names: DOM storage, offline storage, and others.

Though similar to cookies, client-side storage does not have the same size and time limitations as cookies; much more data can be stored and for longer periods of time.

Another key difference is the how the flow of data is controlled. Cookies can be written to the client by the server and retrieved by the server at will. In contrast, information stored at the client must be requested by the client. Likewise, it cannot be read by the server without permission; it must be sent to the server by the client.

The type of data best suited for client-side storage is information that does not change often, like contact information or map locations. One example is the Gmail web app for iPhone and Android, which supports offline data storage (Figure 12-8), allowing you to interact with your mail when you are offline, then syncing with the server once you reconnect.

Figure 12-8. Use of offline storage in the Gmail iPhone web app

Possible applications for client-side storage are as numerous as their potential implementations. The subject of client-side storage is too large to cover completely here, and books will undoubtedly be dedicated to the subject, but know that the methods for reading and writing to it are similar to those methods for reading and writing cookies.

Though WebKit currently supports client-side storage, it is still part of the as-yet-unratified HTML5 proposal and is therefore subject to substantial change before it will become a standard. Already, popular applications that were early adopters of client-side storage have been broken by the changing standard.

CSS

What really makes the iPhone stand apart is its excellent support of CSS and JavaScript. Having desktop-grade CSS support means that you can use the same techniques to create mobile experiences as you would a desktop experience. Not only can the iPhone display a usable version of your site even if it isn't optimized for mobile devices, but you can also create mobile web apps quickly and easily.

Traditionally in mobile, you were lucky if a mobile browser supported some CSS2; in fact you were lucky that most devices supported CSS-MP, the subset of CSS2 meant for mobile devices.

The importance of being able to create a consistent user experience across multiple devices was obviously important, but all but a few browsers were merely paying lip service to CSS support for years. That is, that was the case until the iPhone, which motivated the entire mobile industry to invest in their browsers and their ability to bring CSS-based, web standard designs to mobile devices on par with what we are accustomed to on the desktop web.

CSS2

The iPhone has excellent CSS2 support for a mobile browser. In fact, the iPhone might render CSS a bit better than the desktop web browser you're using these days. Though WebKit and Safari for the desktop support the full CSS2 specification, passing the CSS2 Acid2 test with a 100 percent score (Figure 12-9), Mobile Safari fails on a number of tests, as seen in the following image.

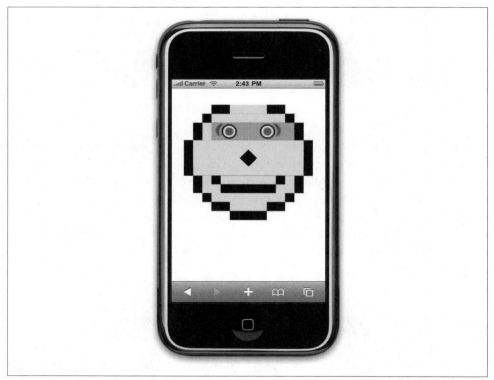

Figure 12-9. How the iPhone fares with the Web Standards Project's Acid2 test for CSS2 support

Though Mobile Safari might not have a perfect score, it is certainly at the head of the class in mobile CSS2 support. The vast majority of styling techniques we employ to create desktop designs can also be used to create our mobile designs.

Positioning and page flow

The iPhone supports the majority of positioning techniques, including relative, absolute and fixed positioning. I say "the majority," as the iPhone currently does not support fixed positioning to the bottom of the perceived viewport, which is useful for creating bottom tabbed navigation. This is because Mobile Safari treats the entire browser as the viewport, so when you scroll the page with your finger, you scroll the entire viewport, including any objects that are in a fixed position, to the bottom of the screen.

In my opinion, this is something that Apple needs to fix soon in order to create more useful mobile web app experiences. Luckily, not all mobile browsers based on WebKit follow this model.

In addition to positioning, page flow styling tools like `display:` and `float:` all work as expected and can be reliably used to create complex styled sites and web apps.

Image replacement

One of the most common techniques in standards-based designs that is still largely missing from mobile design is the ability to replace text with an image defined by your stylesheet. This allows you to maintain a tight separation from your content and presentation, but more importantly, it means that you can load alternative images that are dependent on the stylesheet being loaded—a trick that can come in handy if you plan to support multiple mobile designs from one markup source.

One of the more common examples is replacing the h1 element, often indicating the title of the page, with your logo. For example, you can put your company name as text within the h1 element in your HTML:

```
<h1 id="logo">My Company Name<h1>
```

In CSS, you use the background element to set the logo as a background image, while pushing the actual text beneath the visible area that you define:

```
#logo {
  background: url(logo.png) no-repeat;
  width: 200px;
  height: 75px;
  overflow: hidden;
  line-height: 10em;
}
```

But when it really gets fun is when you want to make the iPhone web app backward-compatible with lower-end devices that do not support the CSS2 styling techniques. In the example of the logo, simply point to the lowest common denominator (LCD) logo instead of the text:

```
<h1 id="logo-id001"><img src="lcd-logo.png" alt="My Company Name" /><h1>
```

In addition to the previous style, when you add the following, you hide the LCD image from iPhones and other capable browsers:

```
#logo img {
  display: none;
}
```

It can be a bit challenging to think of your page in multiple modes, considering how it will be viewed on a variety of devices, but once you get the hang of it, image replacement techniques work well to provide a higher-end experience to the iPhone and its equals while maintaining a simple fallback design for those devices that don't quite cut it.

CSS3

With most mobile browsers, you need to use lots of images and even tables to create simple visual elements like rounded corners, shadows, and semi-opaque areas. Images add kilobytes, which add time and costs for the user to view, and it is our job to create the fastest and lightest experience possible. This is where CSS3 can come in as a wonderful tool for creating complex designs using the minimum of images, making it ideal for mobile design.

The iPhone supports the majority of the CSS3 specification, allowing us to create visually stunning and bandwidth-friendly designs using minimal amounts of code. For CSS3, Mobile Safari 3.0 scores 97 percent in the Acid3 test (Figure 12-10)—nearly a perfect score, and one that certainly makes it a leader among the Class A browsers.

The -webkit prefix

There are a number of CSS3 modules that are being proposed by the W3C, each in various states of recommendation status (check *http://www.w3.org/Style/CSS/current -work#CSS3* for the most up-to-date status). Until all the modules reach candidate recommendation status, it is up to the browser vendors to decide how to implement each module. Typically, what the browser makers do is add a prefix before the proposed style is ratified. In the case of the WebKit, you add a -webkit in front, though other browsers use their own prefixes; in the case of Mozilla, you use -moz and Opera, -o. Though redundant, to be safe, I recommend including them all in your stylesheet, including the proposed standard. For example, to add border radius, use the following:

```
border-radius: 3px;
-webkit-border-radius: 3px;
-moz-border-radius: 3px;
```

I typically start with, say, -webkit-border-radius at design time, then go back through and add the other prefixes once I'm satisfied with the presentation. Another technique I use is to create a class for each CSS3 effect. In the case of rounded corners, I might use <div class="rounded">, then in my CSS attach all the various prefixes to the class .rounded. This gives me a single style declaration to modify in my CSS as CSS3 comes closer to recommendation status.

For an up-to-date status on each of the functions and features of CSS3 and a complete list of prefixes used, check *http://www.css3.info/preview/*.

Figure 12-10. The Web Standards Project's Acid3 test on the iPhone

Selectors

As discussed in the previous chapter, the iPhone, as a Class A browser, supports the majority of CSS attribute selectors. The site CSS3.info has a handy CSS3 selector test that can be performed on mobile browsers, with Mobile Safari 3.0 passing all 43 tests.

CSS3 attribute selectors come in very handy when creating iPhone web apps, as they allow you to target the first or last items in a list or use advanced attribute selectors to target specific elements within your design. Use of these selectors means less markup and CSS code, reducing your overall page weight.

Multiple background images

Multiple background images, when combined with CSS3, are incredibly useful for mobile web apps. For example, the icon created in Figure 12-11 has small 8-bit PNG file with added rounded corners. There is a 24-bit transparent PNG overlay added to create a glare effect. The final result is an icon that is a fraction of the overall weight of using a single 24-bit transparent PNG file. And because the image used for the glare is stored in memory, it can be used over and over again without calling any additional files, as shown in the image to the right of Figure 12-11.

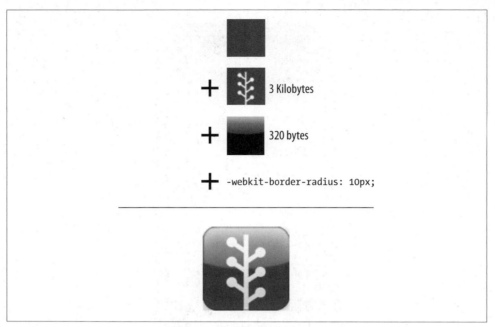

Figure 12-11. Example of using multiple backgrounds for bandwidth-friendly design

Box model sizing

As discussed in the previous chapter, the box model is made up of five areas: the content area, the padding, the border, the margin, and the outer edge. As anyone who has ever tried to create a fluid design knows, creating multiple columns of content with percentage values and adding borders or padding to them is very difficult. For just this reason, box sizing—a CSS3 property supported by the iPhone using the `-webkit` prefix—has been introduced. Normally, when you define a width value for an element that is for the content area, adding borders and padding increases the overall width of the element.

In the following example, a 200-pixel box with a 5-pixel border and 10 pixels of padding would actually render a box 230 pixels wide:

```
<div style="width: 200px; border: 5px black solid; padding: 10px;">
Box
</div>
```

But if you add the `box-sizing` property, you can tell the browser how the width is to be interpreted, as shown in this example:

```
<div style="width: 200px; border: 5px black solid; padding: 10px; -webkit-box-
sizing: border-box;">
Box
</div>
```

The available options are:

`box-sizing: border-box;`
    The specified width is from border edge to border edge

`box-sizing: padding-box;`
    The specified width is from padding edge to padding edge

`box-sizing: content-box;`
    The default method

Box shadow

The `box-shadow` property applies a shadow to the desired element. Like most CSS3 properties, it is useful for reducing the need to use weighty images to achieve desired visual effects:

```
-webkit-box-shadow: hoff voff blur color;
```

Rounded corners

Rounded box corners are a common visual effect used on the Web, but techniques used for creating the effect can add additional markup and images to your code, increasing page weight and reducing compatibility across multiple devices. Using the CSS3 `border-radius` property allows us to easily define rounded borders using nothing more than CSS:

```
-webkit-border-radius: 5px;
```

Additionally, you can specify the specific corner you wish to define, to create a number of different visual effects:

```
-webkit-border-bottom-left-radius
-webkit-border-bottom-right-radius
-webkit-border-top-left-radius
-webkit-border-top-right-radius
```

Border images

The ability to define border images can come in handy to define visually unique controls and elements using a minimal amount of resources:

```
-webkit-border-image: url("border.png") 20 14 round stretch;
```

In the iPhone user interface library, iUI, the border image is used to create buttons using small reusable images, as shown in the Search button (Figure 12-12).

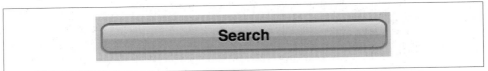

Figure 12-12. The iUI example of a border image

To accomplish this technique, the base image is used as the asset to create the border. The `border-image` property looks at the base image and divides it into nine parts—the four corners, the four sides, and the center (Figure 12-13):

```
border-width: 0 12px;
-webkit-border-image: url(whiteButton.png) 0 12 0 12 stretch stretch;
```

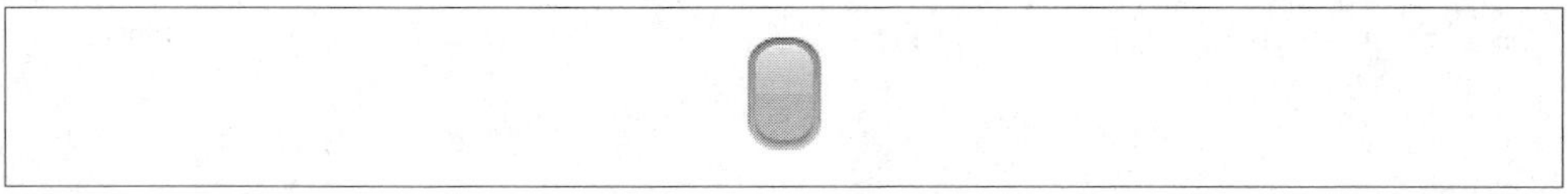

Figure 12-13. The iUI base image used for the border images

In this example, a button is "drawn" around the text in the hyperlink. Additional space has been reserved on the left and right of the text by setting the border width to 12 pixels. There hasn't been any space reserved above or below, because the line height is sufficient. Then the "button" is faked in by stretching the "whiteButton" image as a background for the borders. (If a standard background were added, it would appear on top of the border background where they overlap.)

Text effects

Text effects can be applied in order to reduce the dependency on images to achieve a desired visual effect. The most common is `text-shadow`, which does not require the `-webkit` prefix, and which can used to create a small drop shadow, or a bevel effect, common in iPhone designs:

```
text-shadow: 3px 3px 2px #333333;
-webkit-text-fill-color: #0000ff;
-webkit-text-stroke: 1px #000;
```

Another text property that is useful for iPhone web apps is `-webkit-text-size-adjust`, which is the property that scales the text when the orientation of the device is changed. Adding the following code to the body will prevent the text size from being changed when the orientation changes:

```
-webkit-text-size-adjust: none;
```

Text overflow

When designing mobile interfaces for content that isn't entirely within your control, like a dynamic application, content management system, or imported feeds, it is incredibly useful to present a line or two of text and hide the rest:

```
text-overflow: clip;
text-overflow: ellipsis;
```

In CSS3 we have the handy `text-overflow` property, which can either clip the test at a specified width, or truncate the line at the last visible word and insert an ellipsis (...) to indicate the abrupt end of the sentence.

Text overflow doesn't work by itself; it needs to be combined with over properties, as shown in the following example:

```
.truncate {
  white-space: nowrap; (don't wrap the line)
  width: 200px; (define the visible area)
  overflow: hidden; (hide text outside the visible area)
  text-overflow: ellipsis; (add the ellipsis)
}
```

This property does not require the -webkit prefix.

Visual Effects

There are a number of visual effects that you can perform specifically with the iPhone and iPod touch. These effects are part of the WebKit distribution, but not all devices that use WebKit support them—at least, not yet. This is largely due to how the iPhone is able to offload visual effects to the GPU efficiently, something that is a bit more of a challenge with platforms that are loaded onto licensed hardware, as in the case of Android devices.

Gradients

WebKit supports two extensions to CSS for producing gradients. When added to the background element, the linear and radial gradients can be used to add sophisticated shading and simulate 3-D effects on buttons, and so on, without the use of images. For the prototype, the extension is as follows:

```
-webkit-gradient(type, start_point, end_point, / stop...)
```

or:

```
-webkit-gradient(type, inner_center, inner_radius, outer_center, outer_radius, /
stop...)
```

where:

type
> Is either linear or radial.

start_ and end_point
> Are the positions for the beginning and ending of the gradient, such as left top or right bottom but can be any combination of left or right and top or bottom.

inner_ and outer_radius
> Specify the size of a radial gradient's beginning and ending circles.

inner_ and outer_center
> Specify the start and end positions of the radial gradient, similar to *start_* and *end_point* for linear gradients.

stop

> Can be a series of *color stops* that specify the colors transitioned to and percentages of the gradient the transition uses, expressed as a float value between 0 and 1.

For example, the following code assigns a CSS class `bw-grad` that sports a background linear gradient that transitions from black at the top to white at the bottom:

```
.bw-grad {
  background: -webkit-gradient(linear, left top, left bottom,
  from(rgba(0,0,0,1)), color-stop(1, #fff));
}
```

Masks

Masks are a powerful extension to CSS that can produce effects such as vignetting, soft edges, glare, and images with rounded corners or irregular boundaries. The general idea is that multiple images can be layered. Each layer can use white, black, or its alpha channel to specify what parts of the back image show through the foreground image. If you are familiar with masking layers in Photoshop, you have an understanding of the capabilities of CSS masks.

A simple example is shown in Figure 12-14:

```
<img src="penny.png" style="-webkit-mask-box-image: url(heart.png);">
```

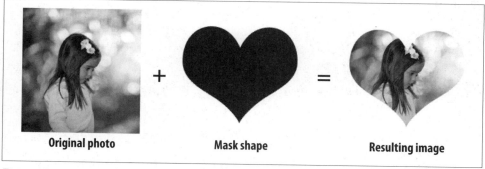

Original photo   **Mask shape**   **Resulting image**

Figure 12-14. An example of using a CSS mask on an image

Transitions

Transitions are an extension to WebKit that can be used to modify a CSS property, such as height or width, over time. Not all properties can be animated with a transition, but the list is fairly large.

The prototype is:

```
-webkit-transition:property duration timing_function delay
```

where:

property
>   Is the name of the property in the same style that is to be modified.

duration
>   Specifies how long the transition will take, specified in seconds or milliseconds.

timing_function
>   Specifies a predefined function on how the intermediate values of *property* are calculated. Valid values are `ease`, `linear`, `ease-in`, `ease-out`, and `ease-in-out`.

delay
>   Specifies how long of a wait there is before the transition begins, specified in seconds or milliseconds.

After the transition has been defined and assigned to an HTML element, when a new value for *property* is assigned (through JavaScript, for instance), the new value isn't immediately assigned. Instead, it transitions to the new value as defined by `-webkit-transition`.

Transforms

Transforms are used for modifying geometry of objects through mathematical operations. They can be used to create visually interesting effects and animations that can be made to appear in a 3-D space. This simpler example rotates all `rotate-me` class elements 30 degrees clockwise (Figure 12-15):

```
.rotate-me {-webkit-transform: rotate(45deg);
```

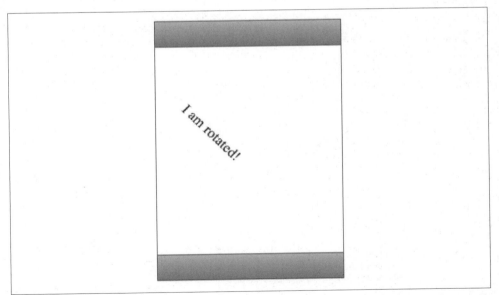

Figure 12-15. An example of CSS transforms in the iPhone, from the Apple Safari Visual Effects Guide

You can also apply multiple transforms:

```
-webkit-transform:translate(-10px,-20px) scale(2) rotate(45deg) translate(5px,10px)
```

Transforms can be used to do incredibly interesting things with your design. In some cases, Apple has created shortcuts to common transform actions to enhance the user experience. Such is the case with Back Face Visibility, which allows you to have a hidden "back" to your content. When the user initiates the transform, it animates to reveal the back face. As documented in this example from Apple:

```
/* Styles the card and hides its "back side" when the card is flipped */
.face
{
  position: absolute;
  height: 300px;
  width: 200px;
  /* Give a round layout to the card */
  -webkit-border-radius: 10px;
  /* Drop shadow around the card */
  -webkit-box-shadow: 0px 2px 6px rgba(0, 0, 0, 0.5);
  /* Make sure that users will not be able to select anything on the card. */
  /* We create the card by stacking two div elements at the exact same location.
     The back of the card is shown when we rotated the card 180 degrees along the
     y-axis. Setting this property to hidden ensures that the "back side" is hidden
     when the card is flipped.
  */
  -webkit-backface-visibility: hidden;
}
```

Animations

Animations, similar to transitions, modify properties over time. However, transitions are one way (from one value to the next), but animations can provide any number of intermediate values that are not necessarily linear. One example is a bouncing box. The box could go up, slow down to a stop, reverse direction, and repeat the process when it reaches the original position:

```
<style>
top:
.bounce {
 -webkit-animation-name: bounce;
 -webkit-animation-duration: 4s;
 -webkit-animation-iteration-count: 3;
}
</style>

<script type="text/javascript" charset="utf-8">
function restartBounce(element)
{
 element.style.webkitAnimationName = '';
 window.setTimeout(function() {
  element.style.webkitAnimationName = 'bounce';
 }, 0);
}
```

```
</script>

<body>
<div class="bounce" onclick="restartBounce(this)">
</div>
</body>
```

Hover, clicks, and taps

The user interface on the iPhone does not include a mouse, so certain behaviors must change. Key among these changes is hovering and clicking behaviors. Because there is no mouse, hovering cannot produce more information, such as status bar updates and tool tips. However, by cleverly using taps instead, the information can still be related to users.

Take, for example, a desktop web application that supports a JavaScript menu. When the user clicks the menu, it drops down with more options. Hovering over these options causes a description for that item to appear in an info box nearby. Clicking an item from the options takes the user to another page. That mechanism doesn't work with the iPhone, because there is no hovering.

Instead, the interface can be modified for the iPhone user so that the description appears in the info box when an item is first tapped, and a second tap takes the user to the selected page.

WebKit supplies an extension to CSS to make these sorts of changes feel natural: tap, highlight. It can be used to cause the tapped element to be all or partially obscured by color. It supports the full range of CSS colors, including alpha:

```
-webkit-tap-highlight-color: blue;
```

JavaScript

The scripting language supported by all major browsers is JavaScript. In an unfortunate naming accident, JavaScript has often been mistakenly considered to be Java's younger sibling, but this is far from the truth, as the two languages have completely different pedigrees. JavaScript shares much with Microsoft VBScript, but is more ubiquitous across browser platforms.

JavaScript is entirely client-side: all the code is run on the browser's computer, not on the server. JavaScript functions can be included from external files, stored in the document's header element, and inline with the HTML. These functions are "fired" on the client at specific times during the page's lifecycle (such as when the document has been loaded), by timers, or via user interaction (such as clicking and hovering).

Although rather humble in this respect, JavaScript is also rather powerful, and is the foundation for a variety of other technologies that power Web 2.0.

DHTML

Of these other technologies, the most obvious (and most simple) is Dynamic HTML, or DHTML. It is a technique, rather than a standard, for modifying the content of the web page and providing an application-like look and feel. Simply put, DHTML uses JavaScript to modify page elements dynamically. JavaScript functions modify either the styles or the HTML elements.

The most common is show/hide, which allows you to hide content using the CSS `display: none;` property; then, when the user touches the element, it changes to `display: block;`. This DHTML technique is incredibly useful when designing iPhone web apps, and has at least one conceivable use in just about every web app:

```
<script type="text/javascript">
<!--
    function toggle_visibility(id) {
      var e = document.getElementById(id);
      if(e.style.display == 'block')
        e.style.display = 'none';
      else
        e.style.display = 'block';
    }
//-->
</script>
```

There is a plethora of user interface libraries that use JavaScript to modify page elements: jQuery, Prototype, and MooTools, to name a few.

Ajax

Making the web application feel lively is half the battle. In most cases, it also has to do something with the data displayed on the page. Enter Asynchronous JavaScript and XML, or Ajax.

With Ajax, the developer can cause events to happen on the server, such as updating records in a database, without causing the page to be refreshed in the browser. This leads to web applications that feel much more responsive. Just like all JavaScript, Ajax can be fired from a variety of conditions, but usually from user actions. An auto-complete search box is a good example. On each keypress, the list of choices is narrowed and updated.

In general, Ajax has three parts: the data sent to the server, the function that is supposed to be performed on the data in the form of a URL, and a request for a response.

The data can be sent to the server in a variety of formats, including plain text, XML, and JavaScript Object Notation (or JSON). The request for a response usually includes the name of a JavaScript function in the web page to be called, conveniently referred to as the *callback function*. This function can then modify the web page based on the data that is returned from the server.

The data and callback function are packaged together in a standardized XML format, called the Ajax request, and sent to the server. The server eventually (remember that this is an asynchronous procedure) returns data to the browser along with a reference to the callback function. The callback function interprets the data and updates the web page in-place without reloading it.

For instance, you may see a "More" link at the end of a list. Clicking "More" calls a JavaScript function that sends a request to the server for another 10 items to display and `addItemsToList`, the name of a JavaScript function in the web page. The server POSTs the response back, which includes `addItemsToList` as the callback function.

Multitouch

The iPhone, as of OS 2.0, includes the ability to recognize and handle both multitouch events as well as gestures. This effectively provides Mobile Safari with many of the same user interface abilities that you might find in the native API. Unfortunately, not much attention has been paid to the addition of multitouch, being overshadowed by the iPhone API.

In this example, we combine multitouch events with visual effects to enable dragging:

```
function touchMove(event) {
  event.preventDefault();
  curX = event.targetTouches[0].pageX - startX;
  curY = event.targetTouches[0].pageY - startY;
  event.targetTouches[0].target.style.webkitTransform =
    'translate(' + curX + 'px, ' + curY + 'px)';
}
```

One incredible example comes from Thomas Robinson (*http://tlrobinson.net/blog/ 2008/07/11/multitouch-javascript-virtual-light-table-on-iphone-v20/*); he creates a virtual light table that pulls in Flickr photos, then uses multitouch to allow you to use two fingers to pinch, zoom, rotate, or simply move the images around on the screen (Figure 12-16).

Fixed Footer

I mentioned earlier that one of the things the iPhone doesn't support is the ability to display fixed position content at the bottom of the viewport, because you can scroll the entire viewport. This is incredibly annoying if you want to create a bottom tab bar, similar to the native app convention.

Richard Herrera created a JavaScript file that adapts for this (Figure 12-17) and is available for free for your iPhone web apps (*http://doctyper.com/archives/200808/fixed-posi tioning-on-mobile-safari/*).

Unfortunately, because this script is technically a hack, it can be sluggish to scroll elements, especially if you have many tabs open in Safari. Following Richard's work,

Figure 12-16. An example of multitouch on the iPhone

Matteo Spinelli created another version of a fixed footer scroller that is a bit smaller in file size and performs a bit better (*http://cubiq.org/scrolling-div-on-iphone-ipod-touch/5*).

Creating a Mobile Web App

Prior to the iPhone, if a site was built for mobile browsers, it was simply referred to as a mobile website, or as web content designed to be viewed within a web browser. Few mobile browsers support the complex interactions that are often associated with web applications, or application-like experiences using web technologies. The iPhone wasn't the first mobile browser to enable the creation of mobile web applications, but it certainly has had the greatest impact.

Mobile web applications designed specifically for the iPhone (and for WebKit by proxy) have since exploded into their own category of mobile experience. But the reality is that creating a mobile web app isn't really that much different than creating a normal mobile website.

Figure 12-17. An example of using CSS transforms to create a fixed footer

The following are techniques to create a more application-like experience specifically on the iPhone.

Defining the Viewport

The *viewport* is the area within a browser where content can be seen by the user. On the desktop, the user can resize the browser window and therefore the viewport. In mobile devices, the browser area cannot be resized; therefore, most Class A browsers—including WebKit and Mobile Safari—create a virtual viewport area, adjusting the content to fit within the screen. By default, Mobile Safari assumes a viewport of 980 pixels, the recommended size for desktop sites.

You can change the viewport to the width of the device—on the iPhone, 320 pixels—by adding the following line of code within the head of your document:

```
<meta name="viewport" content="width=device-width">
```

Alternatively, you can define the default scale of your web application to 1.0 using the initial-scale option, and Safari will determine the pixel values, starting at the device width and device height:

```
<meta name="viewport" content="initial-scale=1.0">
```

To disable the ability of the user to scale the application using the pinch and zoom gesture, add the user-scalable attribute:

```
<meta name="viewport" content="initial-scale=1.0, user-scalable=no">
```

Full-Screen Mode

With Mobile Safari, it is possible to run a mobile web application in full-screen mode without the default Safari user interface, or "browser chrome." This is useful for creating an immersive user experience, on par with a native application. Simply add the previous example line of code to the head of your document, and each time it is launched from the Home screen, it will load without the browser chrome (if launched within Safari, it will obviously have the browser chrome):

```
<meta name="apple-mobile-web-app-capable" content="yes" />
```

This feature can be a great way to leverage web apps to create a native-app-like experience; however, it must be used carefully. The loss of the browser chrome means that there is no forward or back controls; therefore, you need to ensure that your web app contains the appropriate controls for the user to interact with your app. Additionally, your web app needs to use DHTML or Ajax to display content to users, not hyperlinked pages. When running in full-screen mode, any hyperlinks will load the full instance of Safari, breaking the desired effect.

 This is an Apple-only technique for now. Until a standard is defined, expect other browsers to emulate this property to capitalize on the large number of iPhone web apps. Other browsers, like Opera Mobile, allow you to run sites in full-screen mode without a defined meta property.

Changing the Status Bar Appearance

In addition, when using apple-mobile-web-app-capable, you can optionally define the status bar style when running web apps in full-screen mode in Mobile Safari. It can be defined as default, black, and black-translucent, which removes the status bar from the page flow, meaning that your content starts at the top of the page, not directly beneath the status bar:

```
<meta name="apple-mobile-web-app-status-bar-style" content="black" />
```

Adding an Icon

When users bookmark a site or web app on their iPhone, iPhone asks them if they would like to place a link to it on their Home Screen; Apple calls these links Web Clips. When the user chooses to create a Web Clip, the iPhone looks for *apple-touch-icon.png* or *apple-touch-icon-precomposed.png* (used in case you don't want the iPhone to add the gloss effect to the icon) at the root of your site. If found, it will use the icon as the default Web Clip icon. Alternatively, you can define the location using the following example code. The Web Clip icon can be any size, but I recommend that it be 57×57 pixels, output as an 8-bit PNG:

```
<link rel="apple-touch-icon" href="images/apple-touch-icon.png"/>
```

Apple isn't the only platform to use `apple-touch-icon` to define an icon for your web apps. Other mobile browsers and web app portals look for the touch icon referenced in the head of your markup. A standardized means of defining a web app icon will emerge soon, but until then, expect mobile browsers to either adopt Apple's model or invent their own.

Web Apps As Native Apps

It is hard to look at mobile technology and not immediately veer toward building a native application. You can check the user's location, play back local media, interact with the user's address book, plus a whole host of other features that tap into the device's potential. When you add that the APIs are well documented, the storefronts are massive, and that you can actually charge people for the app, it can be hard to make a case for creating a mobile web app versus a native app.

Native applications, however, do not solve the greatest problem in mobile, from its inception to today: device fragmentation. Creating applications for multiple platforms is extremely difficult and costly to do. Making an application for a single platform is not a strategy; it is an opportunity.

I certainly can't argue against the advantages that such opportunities can provide. So what if you could create mobile web apps that work on multiple mobile devices, then deploy those across multiple mobile platforms as native applications? Not only could you tap into the device capabilities, but you could charge for it. You could seize the opportunities on multiple platforms, without having to sink your time and resources into any one platform; plus you wouldn't have to deal with multiple frameworks, programming languages, code repositories, testing cycles, and app store submissions, all of which cut into the profitability of your application—could it be true?

It is, and you can. In fact, on platforms like the iPhone or Android you can build some of the easiest applications in under a day, using WebKit. Apple's *Objective-C Programming Guide* (*http://developer.apple.com/iphone*) describes it best:

Many applications need to display web content in windows, whether it's live content on the web, or static files on disk. Some applications are full-featured browsers, but more often applications embed web content as a convenience, as in a custom document system. HTML is the de facto standard representation of documents on the internet, so it's only natural that you will want to display that content without having to launch a web browser for each file or link clicked by the user.

Many of the most downloaded iPhone applications leverage WebKit to render content. Users don't care if it is native or web content—they just care that it works when they want to use it. Relying on web technologies gives you a common framework to use across all clients—mobile and desktop. You can build a WebKit container and load a local or remote web app, basically running an app within an app, or you can integrate your mobile web app views into specific areas of a custom native app.

PhoneGap

A great tool for building native apps from web apps is PhoneGap (*http://www.phonegap .com*), an open source library that enables you to take a mobile web app and compile it into a native app for the iPhone, Android, BlackBerry, and other platforms. You can simply download the PhoneGap code, then open the project in the appropriate framework development environment; for example, to create an iPhone application, open up the *phonegap.xcode* project in Apple's Xcode (Figure 12-18).

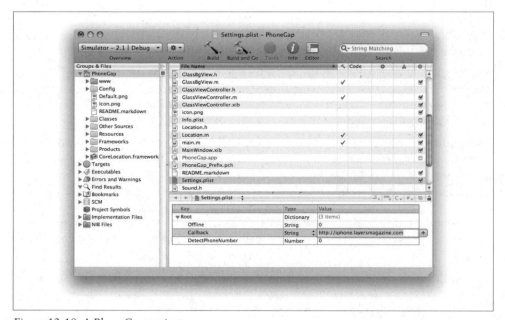

Figure 12-18. A PhoneGap project

You can take an existing mobile web application designed for most Class A browsers and build a native application in minutes. PhoneGap enables you to load an application from both a remote URL—basically, like a web browser—or you can store your HTML, CSS, and JavaScript locally for faster access.

Applications made with PhoneGap occasionally have problems being certified into app stores, but normally with a few tweaks they can get approved.

Tools and Libraries

Many toolkits and interface libraries have emerged to aid in the creation of mobile web apps, specifically for the iPhone. Building on the work of others can be a huge time saver when trying to build a mobile web app.

iPhone GUI PSD

One problem that comes up with creating iPhone web apps is having to recreate the iPhone user interface in HTML and CSS. Even if you are creating your own look and feel, it is still helpful to see Apple's user interface in detail to make your app feel at home on the iPhone. Luckily, the kind folks at Teehan+Lax have provided us with a layered Photoshop file (*http://teehanlax.com/downloads/iPhone_GUI.psd.zip*), complete with all of the iPhone GUI elements (Figure 12-19) needed to create an iPhone-inspired web app, or to be used as the foundation for your own UI.

Figure 12-19. iPhone GUI PSD

iUI

Shortly after the iPhone was released, during the first iPhoneDevCamp, developer Joe Hewitt created an open source user interface library that mimics the appearance and interactions of the iPhone. Originally dubbed iphonenav, the project is now known as iUI (Figure 12-20), and is one of the more popular tools used for creating iPhone web apps.

Figure 12-20. iUI

iUI uses CSS and JavaScript to quickly create menus, animated page transitions, and other effects that make your web app feel at home on the iPhone. It is highly optimized and has a page weight of only 30 KB, making it an ideal toolkit for mobile devices. It uses the single-page model, loading additional content via Ajax into the view, making it incompatible with browsers that don't support JavaScript.

jQTouch

Another interface library called jQTouch is designed to include other WebKit browsers based on the popular JavaScript framework jQuery (Figure 12-21). It was created by designer David Kaneda, out of the need for a lightweight skinnable interface library for more than just the iPhone.

Figure 12-21. jQTouch

Like iUI, jQTouch supports iPhone-style menus, animated page transitions, and effects using CSS transforms. It is nearly four times larger than iUI—around 112 kilobytes, half of which is just the core jQuery library. The flexibility of having a library based on jQuery can be useful, meaning that there is the ability to leverage the simple syntax for creating JavaScript-based web apps as well, and the wealth of jQuery plugins. However, its large file size is certainly an issue to consider before starting to make web apps with jQTouch.

## CHAPTER 13

Adapting to Devices

Not all mobile devices are created equal. Thus the age-old problem in mobile design and development: devices can be vastly different from each other. It would be easy if different devices simply supported different attributes—one supports CSS3 and one doesn't. But it isn't that easy. One device might support CSS3 and another device might support CSS3 poorly—or worse, incorrectly.

Honestly, this might not be a problem at all if we only had a few platforms or browsers to contend with. For example, how many big desktop platforms are there? Three: Windows, Macintosh, and Linux, with the first two making up almost 95 percent of the market. And how many big desktop web browsers are there? Four: Internet Explorer, Firefox, Safari, and Opera, with the first two making up nearly 90 percent of the market, the most recent versions of which all pass the Acid2 test for CSS2 support, effectively making them web-standards-compliant.

The mobile industry is an entirely different story altogether. In mobile, you have a half a dozen or so platforms, like the S60, iPhone, Android, BlackBerry, Windows Mobile, and LiMo smartphone platforms, plus a dozen or so feature phone platforms. Add the plethora of mobile web browsers that run on each of these platforms, for which less than 1 percent of all mobile browsers are able to pass the Acid2 test for CSS2 support, for example, and you can start to see that the mobile web is a very fragmented and difficult space to support.

Case in point: one would assume that if I enter *www.domain.com* into a web browser, I would be taken to the appropriate experience for my device and viewing context, right? This simple action is what we've promised to users regarding how the Web works. But of course in mobile, it isn't that simple. Sometimes entering *www.domain.com* detects your device and sends you to the desired mobile experience. Sometimes we have to create *m.domain.com*, and the recent trend is to also create *iphone.domain.com*.

Where does it end? At what point do we just tell users that when they want to go to a site, they should enter a web address, regardless of their device? This is the topic of this chapter: adapting for multiple devices.

Detecting, adapting, and supporting multiple devices has historically been the worst pain point of mobile design and development. It usually requires a software system to perform a three-step process in order to render the best experience per device, creating plenty of opportunities for problems. Though there are several strategies to solve this problem, each of them work in this basic way:

Detect
> The system must first detect the device requesting the content, known as device *detection*. This typically requires a valid and up-to-date device database, with all the pertinent information on the hundreds of devices on the market, or in some cases a very smart mobile web browser.

Adapt
> It next takes the requested content and formats it properly to the constraints of the device, which is called *content adaptation*. This might include the screen size, device features, appropriately sized photos, supported media types, or web standards support. This is done by having an abstracted layer of content and a layer of presentation and media for each supported device.

Deliver
> Finally, it must serve the content to the requesting device successfully, usually requiring testing each device or class of device that it is intended to support.

This entire process is often referred to as *multiserving*, *device adaptation*, or simply *adaptation*. As you might imagine, this process requires the consideration and adaptation of many variables. It is in no way an obstacle that cannot be surpassed, but it does add significant cost and complexity to a mobile project.

Luca Passani, who maintains the largest open source device database (a necessity in this process), puts it this way:

> Multiserving is a necessity, but creating (and maintaining) more than three versions means going down a very slippery slope.

So what are we, as designers and developers, to do? We know that mobile technology is important—it is the platform for the future, and we want our content in the space—but dealing with all of these devices seems like a headache that we don't exactly want or need.

I've said before that whenever there are multiple options in mobile, there are no right or wrong choices—just what is right for the user. Let me stress that point here again as well. Not only are there no right or wrong answers in how you adapt for different devices, but there is strong support and intense opposition for a number of different approaches. Members of both the web and mobile communities have an almost religious belief that one approach is better than another, either through dogmatic principle or through years of experience.

Even this book isn't immune to this debate. Each of the technical reviewers invited to review this chapter offered different suggestions and techniques for how to solve the problems of adapting for devices. Some were based on how the adaptation should be, and some were based on how to get it done in today's environment. If I, as someone with more than a decade of experience in the mobile ecosystem, struggle to make sense of this topic, then I can only imagine how you must feel. However, I can provide my own perspective on the topic, which is a bit unique.

I come from a family of entrepreneurs. My father was an inventor and business owner. Each of my six brothers (of which I am the youngest) either own or have owned their own businesses. You could easily say that entrepreneurship runs in my family. I'm not immune; I own my business now, but I certainly didn't start there. Like most people, I've worked for a number of different companies, and many of them just happened to be mobile companies.

In fact, I've had the pleasure of working with some really great companies and brands over the course of my career in the mobile industry. This topic, adapting for multiple devices, has been the coffin nail for the majority of them. Not only did the costs run them out of the mobile business altogether and force them to sell off assets way below their value, but the companies also started to lose faith in the long-term viability of the medium.

As an entrepreneur, I cannot in good faith recommend to anyone something that I wouldn't be willing to make myself. To me, this is the only principle that matters. It isn't a matter of industry dogma: it is a matter of business ethics.

In my experience, adapting for multiple devices is the greatest risk that your project can make. Too many challenges are unforeseen, there is too much knowledge that takes too much time to learn, and the hidden costs can too easily multiply beneath you. As you might imagine, I get a bit conflicted, as I also believe that mobile will be the most powerful and defining medium of the next 25 years.

So how does one balance the necessity of multiserving while still keeping mobile investments of time and money under control? I've found five basic strategies to work around this seemingly insurmountable obstacle:

1. You can do nothing and wait for the mobile industry to become fully homogenized.
2. You can try to use a progressive enhancement technique to provide fallback experiences to a number of devices.
3. You can target only a handful of devices that support the standards you wish to support, knowing full well that this means making your mobile experience less accessible to your intended market.
4. You can adapt the experience based on the class of the requesting device, making many assumptions about its capabilities.
5. Provide an experience specific to each requesting device.

In this chapter, I will discuss all five approaches to adapting experiences for multiple devices. I will focus primarily on the mobile web, but there is no reason why most of this knowledge couldn't apply to all mobile experiences as well.

I will try to leave out debate as much as possible and just report what I believe that you need to know. There are of course many elements of this topic that could be expanded upon; in fact, an entire book could easily be written on just this one topic. But for the purposes of this book, I will introduce to you the general principles of adaptation and will leave it up to you to decide what is best for you, and more importantly, what is best for your users.

Let's start with the circumstances that might cause you to consider an adaptation strategy in the first place.

Why Is Adaptation a "Necessity"?

Is multiserving a "necessity" like Luca suggests, and if so, why? Yes, multiserving your content in one way or another is absolutely a necessity. If you have a website and you want to have a mobile website, web app, or even native application that shares content in some way with another context, you are going to need to figure out a strategy to make that happen.

In Figure 13-1, you can see a simple example of the software systems that someone might need if she wanted to have a desktop site, a mobile website, a mobile web app, and a native app. Each context is creating an standalone system, because more often than not, tying them all together would be costly in time and resources.

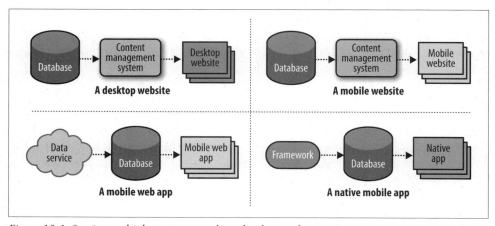

Figure 13-1. Serving multiple contexts can be redundant and expensive

I've found this to be the most common multiserving strategy with most companies I've worked with: creating an internal API or data service to get each system talking to each other, although more often than not the API is actually designed only to pull information from one source into another, and doesn't always send information as well, or at least not as easily. Getting information out of a central database through an API can be a snap, allowing users to create new sessions on the client, but storing data either locally or back into the context-specific database can be a hassle. The problems occur when trying to get each of these separate systems in sync.

For example, say we wanted to create a grocery shopping application, allowing users to create shopping lists from home, then use their mobile device as a shopping list at a physical grocery store. As we really only need to send data one way, it might not be too problematic—a fairly simple multiserving strategy would most likely work. But what if we wanted the application to work in the reverse as well, allowing users to create a shopping list from their mobile device while they are out and about, then when they get home, to use it to place an online order for delivery? The more the user changes his context, the more complex the system needs to be to adapt.

Ideally, we could use one system for multiserving all of our content, as shown in Figure 13-2, with one central source of data outputting content to multiple contexts, using whatever external tools are needed.

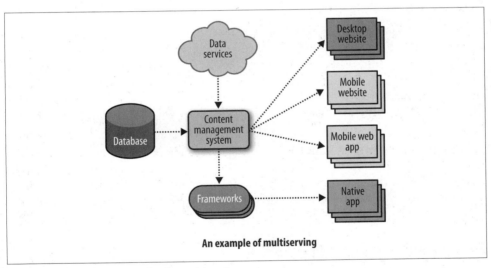

An example of multiserving

Figure 13-2. A simple example of a multiserving system

The question isn't whether it is a necessity; in today's environment, and with the growth of mobile devices starting to move beyond just phones, this isn't a problem that is going away anytime soon. The question is how to adapt for multiple devices with a minimal number of headaches. This brings us to our first philosophical debate: the adaptation/multiserving conundrum, and our first adaptation/multiserving strategy.

Strategy #1: Do Nothing

Our first of four multiserving strategies is to simply do nothing, or rather to wait for the technology to adapt to our principles—which actually isn't as foolish as it might sound.

In 2005, the W3C created the Mobile Web Initiative (MWI) to attempt to bring the standardization it achieved with the desktop web to mobile. This was a difficult challenge, because the W3C had not issued any specifications for mobile standards. The announcement of the effort included an introduction from Tim Berners-Lee about the importance of "One Web," a concept that has been a hotly debated topic in both the web and mobile communities ever since.

Initially, the W3C defined One Web this way:

> One Web means making the same information and services available to users irrespective of the device they are using.

In other words, content should be published once and the device should be smart enough to know what to render; effectively, we wouldn't need to do anything more than ensure that our content meets the content standards. This actually makes a lot of sense, as we'd be able to write once and publish everywhere to multiple devices and multiple contexts.

This do-nothing approach is actually a multiserving system in its own right: it detects devices, though without the need of a device database; it adapts content based on the requested device through thoroughly separated content and presentation and renders it to the requesting device. The multiserving system in this scenario is the browser, assuming that it can intelligently render the appropriate experience that we define for it.

Five Assumptions About One Web

I'll be honest: this approach can work really well when you design your content with it in mind, but it isn't very flexible or robust, and it isn't without flaws:

- It assumes that your content for multiple contexts will be the same, when it usually isn't.

- It assumes that cost per kilobyte to the user is minimal or nonexistent. In other words, it assumes that the user is willing to pay for content not designed for her context.

- It assumes that a persistent and high-speed data network will always be available.
- It assumes that mobile browsers are smart and will support the same standards consistently, which isn't the case, at least today.
- It assumes that a technology-based principle should come before the needs of the user.

These five assumptions will go away in time. In fact, the iPhone and other modern mobile browsers have been designed specifically to address the first four assumptions, which is why Apple has described the iPhone as a "One Web" device.

The problem with the One Web principle is that fifth assumption. It can be dangerous to put the desire for efficient technology before the needs of users. Creating a great experience means finding the sweet spot between the constraints of the technology and the needs of the user.

For example, take the first assumption. Assuming that the needs of a user on a mobile device are the same or even similar to a user on a desktop computer would be foolhardy. Even the needs of a user with a limited phone device are different from those of a user with a smartphone. After you add additional devices, such as GPS units, gaming consoles, or even subnotebooks, you start to see that the features you might include for one device or one context might not apply to another.

Trying to state a principle saying, "It should be done this way" adds risk that people will create poorly adapted experiences. Trying to shoehorn the user into an experience designed for the wrong device for the sake of simple technology or even a high-minded principle is simply irresponsible. The user deserves more and it is our job to provide it.

The One Web Aftermath

After the One Web principle was announced, the mobile community, which had been working on this problem for years, exploded in vehement disagreement over the principle. Many people, including myself, called it unrealistic and said that it ignored the needs of mobile users. Meanwhile, the web community applauded the idea, wanting to see the mobile landscape become more aligned with the Web, and demanding smarter mobile web browsers, thus starting a mild feud that continues to this day.

After working closely with the mobile community, the W3C and the MWI revised the definition, and this is how it still reads today:

> One Web means making, as far as is reasonable, the same information and services available to users irrespective of the device they are using. However, it does not mean that exactly the same information is available in exactly the same representation across all devices. The context of mobile use, device capability variations, bandwidth issues, and mobile network capabilities all affect the representation. Furthermore, some services and information are more suitable for and targeted at particular user contexts.

In other words, content should be published once and the device should be smart enough to know what to render, unless it can't or doesn't make sense to.

Using This Strategy with Media Queries

We've clearly seen in terms of usage with more advanced mobile browsers, such as WebKit or Opera, that when rendering normal web content at desktop quality users, when given the choice, prefer the mobile optimized version designed for the mobile content instead of the often more full-featured desktop version. This actually means that the One Web principle does work if the first four of the assumptions about One Web are true.

This do-nothing One Web approach to multiserving means that you don't need to do anything more than add a line of code that points to a CSS file that defines the presentation for each supported device; called a Media Query, it is part of the CSS3 specification. It requires a higher-end mobile web browser that supports CSS3, like the iPhone, Android, Palm's webOS, or Opera Mobile.

For example, if you wanted to load only a stylesheet for a device that is 320 pixels wide, you could define it this way:

```
<link media="screen and (device-width: 320px)" rel="stylesheet"
href="320styles.css" type="text/css" />
```

However, because many devices can alter their orientation to 480 pixels wide, you would use the following:

```
<link media="only screen and (max-device-width: 480px)" rel="stylesheet"
href="webapp.css" type="text/css" />
```

You can also layer in additional stylesheets for other devices that might fall outside of the traditional mobile device (in this example, devices with screens larger than 481 pixels):

```
<link media="only screen and (min-device-width: 481px)" rel="stylesheet"
href="desktop.css" type="text/css" />
```

You can use fairly complex media queries to fine-tune the presentation for a number of devices, as long as the devices support it.

Strategy #2: Progressive Enhancement

In Chapter 11, I talked about the concept of progressive enhancement, or the technique of using web techniques in a layered fashion to allow anyone with any web browser to access your content, regardless of the browser's capabilities. This is actually our second multiserving strategy.

Though progressive enhancement is normally thought of as a development strategy, it can also be used as an adaptation strategy as well, going through each of the three steps of multiserving. Similar to "doing nothing," this approach assumes that the browser will be smart enough to detect, adapt, and deliver the right experience, but in this strategy we design a site to have several fallback points, supporting a larger number of devices and not requiring support of media queries.

For example, take Figure 13-3, where we can see our presentations layered on one another. At the bottommost level, we have a rich CSS3-based experience for our Class A browsers, including support for media queries. Above it, we have good CSS2 support, for Class B browsers, and so on, until we get to the top level, or our last fallback position of no styling whatsoever, for our Class F browsers. In this case, as long as our markup is semantically coded, it should still be usable on any device that can render HTML.

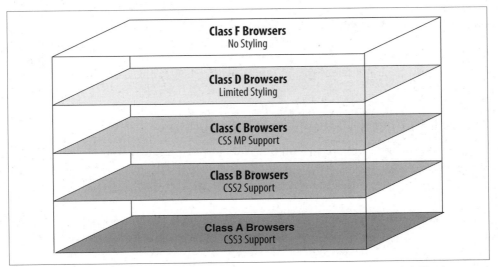

Figure 13-3. Using a progressive enhancement technique to establish several fallbacks to our presentation

The Handheld Media Type

In order to achieve this technique, we are going to rely on the browsers, but luckily there are couple of attributes in XHTML that will help us out. First, let's talk about the media type attributes, which are basically the parents of media queries. XHTML includes a number of media type attributes that you can add to your document to load different stylesheets for different contexts. For example, when we reference our stylesheet, we can define it as a screen, print, projection, or (the unfortunately named) handheld media type. For example:

```
<link media="screen" rel="stylesheet" href="desktop.css" type="text/css" />
<link media="handheld" rel="stylesheet" href="mobile.css" type="text/css" />
```

Or within a stylesheet or the `style` element, we could define it this way:

```
@media handheld {
  * { font-family: sans-serif }
}
```

All desktop browsers are smart enough to know that the screen media type is for them. It's a little-known trick that most mobile browsers have supported the handheld media type for years. This means that we can use the same content and display it on multiple mediums with a single line of code.

However, there are a few catches. First, because the handheld stylesheet will be used on most mobile browsers, the stylesheet should be designed for the lowest common denominator, using basic style techniques and providing limited design opportunities. Mobile browsers that aren't smart enough to support the handheld media type, usually falling into the Class D grouping, could be problematic to support with this approach. The second caveat is that more advanced mobile browsers will load the screen media type instead of the handheld media type because they can render desktop websites reasonably well.

Layering Multiple Stylesheets for Multiple Devices

We can use the media queries technique from our first strategy to target our Class A browsers:

```
<link media="only screen and (max-device-width: 480px)" rel="stylesheet"
href="class-a.css" type="text/css" />
```

This stylesheet would contain our more advanced styling, like rounded corners, heavy use of image replacement, and other advanced styling techniques, and would only be loaded for browsers that support CSS3.

Class B browsers typically, but not always, prefer to support the screen media type instead of the handheld media type. Therefore, we can add another stylesheet for these devices, with fairly basic CSS2 styles, some image replacement, and some placement, but nothing too aggressive:

```
<link media="screen" rel="stylesheet" href="class-b.css" type="text/css" />
```

For Class C and Class D browsers that support the handheld media type, we can then layer in another stylesheet. This time we use a very basic stylesheet. The simpler it is, the better it will render on devices:

```
<link media="handheld" rel="stylesheet" href="class-c.css" type="text/css" />
```

When we add all three of these together, we see something like this:

```
<link media="handheld" rel="stylesheet" href="class-c.css" type="text/css" />
<link media="screen" rel="stylesheet" href="class-b.css" type="text/css" />
<link media="only screen and (max-device-width: 480px)" rel="stylesheet"
href="class-a.css" type="text/css" />
```

The browsers will do their best to load and render the best stylesheet for the browser's capabilities. Now this is hardly a foolproof approach. It involves putting a lot of the burden on the mobile web browsers, which more often than not prove disappointing. However, this approach does work fairly well to reach most of the top devices accessing the mobile web today.

This adaptation strategy isn't the most popular approach with many in the mobile community, as it doesn't give you a lot of opportunities for fine-tuning your styles to the device. Imagine you have a popular browser that is actually quite capable but keeps loading the handheld stylesheet instead of your screen stylesheet. There isn't a lot that you can do to account for these oddities. This strategy can also present a user experience concern, in that you aren't providing the user with the ability to choose which experience he would prefer to load, as recommended by many mobile advocates.

I find this approach works extremely well for simple mobile web experiences, though in the interest of full disclosure, some of my technical reviewers discourage this approach, saying it is too unreliable. I think it all comes down to an issue of what browsers you plan to support. If you want to support the top browsers but still offer something for people with less-capable devices, this strategy might be the one for you. You certainly can't support everyone with it, but it is a quick and easy first step.

Strategy #3: Device Targeting

The third multiserving strategy is *device targeting*, or calling out specific devices by class or model and delivering an experience designed with that device in mind. In this strategy, we don't assume that browsers are trustworthy enough to get to where they need to be. Therefore, in this strategy, the first step of multiserving is to reliably detect the device.

Once the device and browser are detected, you route the user to the best experience. This can be done by checking the HTTP headers, starting with the User-Agent string, in order to recognize the device and browser and then deliver a device-specific site.

One of the more common examples here would be wanting to treat a device like the iPhone differently than the rest of your mobile devices; the markup and styles used to create the iPhone experience could be quite different than what you might use for all the other mobile devices. In these cases, each experience is often built as a standalone product, with little to no adaptation taking place.

Though you could certainly perform dynamic adaptation—something I will discuss more in the next strategy—in this strategy, it is common for developers to utilize a more simplistic approach, relying on the content management system or web application framework to create multiple experiences from a single data source. Create specific experiences for targeted devices, then route devices accordingly.

Because device detection is such a crucial component of this strategy, let's first discuss some of the challenges associated with this step.

The Device Detection Dilemma

Device detection has been a technical challenge for many years. Although it's hardly a challenge to those experienced in mobile development, I feel that the mobile community severely underappreciates how much of a challenge this is for the rest of us.

Large publishers with experienced developers can certainly hack their web server configurations to detect and route devices, even though it still requires having an up-to-date device database with all the device profiles and appropriate techniques to recognize a device. And in larger organizations it can take months to prepare, test, and QA any major changes to the server directives. For smaller companies using shared hosting, altering the server configuration at the levels needed isn't even an option.

In my experience, only a handful of companies that I worked with ever employed a server-side device detection strategy. The vast majority of them were mobile companies whose business was to provide excellent mobile experiences to lots of devices.

This is obviously less than ideal. For the mobile community and the mobile web to flourish, these tools need to be available to all for a price everyone can afford. For many years, I've been arguing that device detection is a problem that we need to solve. The need to provide some level of adaptation will always exist in one way or another, and there will always be many devices and many contexts to which we will want to adapt. However, as mobile device and browser makers adhere more closely to the defined and accepted standards, the variation required will become almost inconsequential.

Device fragmentation will go away one day, but device detection will not. We will always require a method of easily routing experiences for the user. Ironically, this is actually the easiest problem to solve. It just takes people with the experience and commitment to make it happen.

Luckily, there are a few people who solved this problem already, so we don't have to.

Andy Moore's Mobile Browser Detection

One of the more popular solutions to provide simple detection is the PHP script written by Andy Moore (*http://www.detectmobilebrowsers.mobi*). Based on the WordPress Mobile Plugin, which he also wrote, this script looks at the requesting user agent string and abstracts the data to match against the majority of mobile devices, then routes it to the mobile version of your site. All you need to do is include the file with the function, then call the function before your PHP pages do anything else:

```
include('mobile_device_detect.php');
mobile_device_detect();
```

The script is easy to install, works as advertised, is free for personal use, and is available for a small fee for commercial applications. Of course your application needs to be written in PHP for it to work, but for many smaller sites or web apps looking to do simple device detection, Andy's script is the way to go.

Greg Mulmash's Mobile Browser Detection

Greg's script (*http://www.brainhandles.com/techno-thoughts/detecting-mobile-brows ers*), also written in PHP, goes one step further. Like Andy's script, it looks at the requesting user agent string and looks for data recognizable as a mobile device or browser. Using the WURFL database of 6,750 different mobile browser User Agent IDs, this script caught 94.34 percent of them.

Simply create a PHP document with the following code in it:

```php
function checkmobile(){

if(isset($_SERVER["HTTP_X_WAP_PROFILE"])) return true;

if(preg_match("/wap\.|\.wap/i",$_SERVER["HTTP_ACCEPT"])) return true;

if(isset($_SERVER["HTTP_USER_AGENT"])){

// Ignore iPhone and iPod touches

if(preg_match("/iphone/i",$_SERVER["HTTP_USER_AGENT"])) return false;

// Quick Array to kill out matches in the user agent

// that might cause false positives

$badmatches = array("Creative\
AutoUpdate","OfficeLiveConnector","MSIE\ 8\.0");

foreach($badmatches as
$badstring){if(preg_match("/".$badstring."/i",$_SERVER["HTTP_USER_AGENT"])) return
false;
}

// Now we'll go for positive matches

if(preg_match("/Creative\
AutoUpdate/i",$_SERVER["HTTP_USER_AGENT"])) return false;

$uamatches = array("midp", "j2me", "avantg", "docomo", "novarra", "palmos",
"palmsource", "240x320", "opwv", "chtml", "pda", "windows\ ce", "mmp\/",
"blackberry", "mib\/", "symbian", "wireless", "nokia", "hand", "mobi", "phone",
"cdm", "up\.b", "audio", "SIE\-", "SEC\-", "samsung", "HTC", "mot\-", "mitsu",
"sagem", "sony", "alcatel", "lg", "erics", "vx", "NEC", "philips", "mmm", "xx",
"panasonic", "sharp", "wap", "sch", "rover", "pocket", "benq", "java", "pt", "pg",
"vox", "amoi", "bird", "compal", "kg", "voda", "sany", "kdd", "dbt", "sendo",
"sgh", "gradi", "jb", "\d\d\di", "moto");

foreach($uamatches as $uastring){
if(preg_match("/".$uastring."/i",$_SERVER["HTTP_USER_AGENT"]))
return true;
}

}
```

```
    return false;
    }
```

Once this is in place, you can check for a mobile browser and, if it returns true, conditionally load your stylesheet. This technique is more reliable than using a media type, as only one stylesheet will be presented to mobile browsers:

```
<!DOCTYPE html PUBLIC "-//W3C//DTD XHTML 1.0 Transitional//EN"
    "http://www.w3.org/TR/xhtml1/DTD/xhtml1-transitional.dtd">
<html>
<head>
<title>My Page Title</title>
<?phpif(checkmobile()){
echo "<link rel=\"stylesheet\" type=\"text/css\" href=\"/mobilestyle.css\">";
} else {
echo "<link rel=\"stylesheet\" type=\"text/css\" href=\"/regularstyle.css\">";
}
?>
<meta http-equiv="Content-Type" content="text/html; charset=utf-8"/>
...
```

Or as with Andy's script, you can also just redirect traffic to a mobile site that lives in an entirely different location:

```
<?php
if(checkmobile()) header("Location:http://m.domain.com");
?>
<!DOCTYPE html PUBLIC "-//W3C//DTD XHTML 1.0 Transitional//EN"
    "http://www.w3.org/TR/xhtml1/DTD/xhtml1-transitional.dtd">
<html>
...
```

This script is completely free to use, but like Andy's script, requires that your code be in PHP.

The Switcher

The Switcher is a device detection layer to route mobile devices to mobile experiences, available in Java, PHP and .NET, provided by WURFL maintainer Luca Passani.

Whenever an HTTP request is received, the Switcher will analyze each header, checking to see whether it is a mobile device or browser. If it is, the Switcher will direct the requesting client to the URL that's most appropriate for it. It provides a great deal of customization to suit your specific needs.

It is available to purchase as full source code from *http://www.passani.it/switcher/* for €500.

htaccess-Based Device Detection

Another option available to sites running on Apache-based web servers is to create an *.htaccess* file to perform a device lookup method similar to the previous solutions. *.htaccess* files are magic little files that allow you to do URL rewriting, redirection, and other handy tricks that help to provide a positive user experience.

This example is provided by Ryan Neudorf (*http://ohryan.ca/blog/2009/02/18/revisiting-mobile-redirection-using-htaccess-rewrite-rules/*):

```
# don't apply the rules if you're already in the mobile directory
# infinite loop
# you'll want to test against the host if you're using a subdomain

RewriteCond %{REQUEST_URI} !^/mobiledirectoryhere/.*$

# if the browser accepts these mime-types,
# it's definitely mobile, or pretending to be

RewriteCond %{HTTP_ACCEPT} "text\/vnd\.wap\.wml|application\/vnd\.wap\.xhtml\+xml"
[NC,OR]

# a bunch of user agent tests
RewriteCond %{HTTP_USER_AGENT} "sony|symbian|nokia|samsung|mobile|windows
ce|epoc|opera" [NC,OR]
RewriteCond %{HTTP_USER_AGENT} "mini|nitro|j2me|midp-|cldc-
|netfront|mot|up\.browser|up\.link|audiovox"[NC,OR]
RewriteCond %{HTTP_USER_AGENT}
"blackberry|ericsson,|panasonic|philips|sanyo|sharp|sie-"[NC,OR]
RewriteCond %{HTTP_USER_AGENT}
"portalmmm|blazer|avantgo|danger|palm|series60|palmsource|pocketpc"[NC,OR]
RewriteCond %{HTTP_USER_AGENT} "smartphone|rover|ipaq|au-
mic,|alcatel|ericy|vodafone\/|wap1\.|wap2\.|iPhone|android"[NC]
```

JavaScript-Based Device Detection

Yet another way to detect higher-end devices that support JavaScript is to look up the user agent and redirect if the value is true. For example, if we wanted to detect iPhones or iPod touches, we might add the following to the initial page of our domain:

```
<script type="text/javascript" charset="utf-8">
        if (navigator.userAgent.match(/AppleWebKit/i) &&
navigator.userAgent.match(/Mobile/i)) {
            window.location.replace('/path/to/iphone/site/');
        }
</script>
```

Although this approach is the easiest to add to your site, it only works with devices that support JavaScript, which aren't many. It gets tricky when you're trying to support multiple devices, and requires the mobile device to load the desktop content first before the redirect occurs, which costs the user time and money.

Reverse Device Detection

I suggested an alternative approach to the problem in an article I wrote several years ago; I call this alternative *reverse device detection*. In this technique, the primary content on your domain is your basic or lowest-common-denominator mobile content. This ensures that basic devices, such as low-end mobile devices, eBook readers, or other basic mobile devices, still have a usable experience.

Use JavaScript to detect the screen size, looking for screens at least 800 pixels wide, then redirect those users to the desktop site. Although it is not a popular idea to not have your primary site live at the root of your domain, this technique is actually quite efficient, because desktop browsers, or even high-end mobile browsers, are far fewer in number, more consistent, and much easier to detect.

Depending on how similar your mobile site architecture is to your desktop architecture, this technique can also have benefits such as making your site more accessible to those who use assistive devices as well as making your content more search engine friendly.

WordPress Mobile Plugin

As mentioned before, Andy Moore developed a plugin for the WordPress blogging tool, which is very popular and free. As an increasing number of people start to use Word-Press for purposes beyond personal blogging, this plugin makes a simple and easy way to do device targeting. This script not only does the detection, but also provides light adapting as well.

You can see how simple it is from Andy's instructions:

1. Download the plugin (*http://wordpressmobile.mobi/download.zip*).
2. Unzip the download.
3. Upload *wordpress-mobile.php* to your *wp-content/plugins* folder.
4. Activate the plugin.
5. Configure the plugin to your needs.

That's it. Follow those steps and your WordPress site will automatically detect mobile devices and show them a mobile-ready version more suited to a small screen.

dotMobi WordPress Mobile Pack

James Pearce and the folks at dotMobi have taken WordPress integration one step further, turning WordPress into a full-fledged mobile content management system. Their solution provides the following features:

- Mobile switcher to detect mobile visitors and provide an appropriate experience
- Base mobile theme for quick-and-easy XHTML-MP compliance
- Extended mobile themes so that you can unleash your mobile creativity
- Transcoding and device adaptation to optimize the mobile experience
- DeviceAtlas integration for world-class adaptation
- Mobile admin panel for when posts can't wait

- Mobile ad widget to make you some money
- Barcode widget to help users bookmark your blog

The dotMobi WordPress Mobile Pack (*http://mobiforge.com/running/story/the-dotmobi -wordpress-mobile-pack*) is available for free.

Mobile Fu

Moving away from PHP-based solutions, Mobile Fu is a free plugin for Ruby on Rails (*http://github.com/brendanlim/mobile-fu/tree/master*). It will automatically detect mobile devices that access your Rails application, then dynamically define the stylesheet to be viewed depending on the requesting device.

And Many More...

This is just the beginning; new solutions to the problems of device targeting are emerging every month. Content management systems and web application frameworks are incorporating mobile templates and logic into their core. I imagine that one day soon, you won't be able to run a web-based publishing system without the ability to output a mobile version and route devices to it, built right in.

Strategy #4: Full Adaptation

The fourth and final multiserving strategy is *full content adaptation*, or the process of making extremely unique mobile experiences based on the device that is requesting the content—almost always dynamically. Like the other multiserving strategies, it starts with detecting the device that is requesting content and matching that to a valid user-agent string; then the system outputs markup, styles, and images generated exclusively for that device.

Let's say, for example, that we want to support 20 devices across multiple operators. These devices carry browsers that range from Class A to Class D, each with different screen sizes and device capabilities. Using a full adaptation strategy, we detect, adapt, and render, but then we take it a step further. For each request, we detect each of those device's user-agent strings against our device database and then dynamically adapt our base templates to suit that device class and render to the device.

With full adaptation, we dynamically create four specific experiences for each of our classes based on a number of templates and assets designed to degrade or adapt that would render a unique experience to each device.

In addition, we may have interclass optimizations. An example of this might be when you have the same device deployed on two different operator networks. This is a more common problem in the U.S. market, where we have more than just one type of network standard. Occasionally, multiple devices can share the same model number and even have similar user agents but render content differently. In this case, we need full adaptation to do detailed lookups and make sure that we detect the right device on the right network, then do interclass optimizations for that particular device.

How complex people perceive full adaptation to be will vary based on their development experience. I often stress how difficult adaptation is, and then a developer will always come up and offer a simple solution to solve it: "Couldn't you just rig the so-and-so to detect the doohickey, then render the whatchamacallit?" And every time my answer is the same: "You are exactly right. That would work." I pause so the developer can be proud of his inventive solution before I deal the soul-crushing blow of saying, "Now multiply that by the devices you plan to support."

Although content adaptation by itself is a relatively easy problem to solve, the challenge occurs when you multiply your solution by 10 or 20 devices, which are considered by most to be a modest regional launch. For every device you adapt for, you now have to support it, test it, and adjust code to account for it. Content adaptation can easily add an order of magnitude in complexity for every device you support. This of course quickly leads to cost overruns, reduced margins, and a consumption of valuable resources.

Working "On Deck"

The publishers that employ full adaptation are usually the ones working with operators to put their content on the operators' mobile portal, often referred to as a "deck." At this point, we haven't discussed much about the last part of multiserving: delivery. To ensure that your detection and adaptation system works, you need to test and confirm that it is finally delivered to the device successfully. With most multiserving strategies, you can typically test anywhere from 10–20 different devices and cover all your bases. When you work with operators, this is a very different picture.

Operators often demand that in order for publishers to be on their deck, they must support two to three years' worth of devices, which can easily be 20–40 different devices offered each year by each operator. At the very least, this means supporting 40–80 different devices, and at most, 60–120 devices for a single operator.

To put this into perspective, most markets around the world might have only two or three major or Tier-1 operators, so the total number of devices you might need to support is 40–80 device models, depending on the operator agreements. Plus, they all operate on the same GSM network standard, so you rarely have to test the same device twice. In the U.S. market, we have four Tier-1 operators, and as many Tier-2 or regional operators, and several different networks. So you might have 20–40 device models per operator, but because each of those devices is provisioned a bit differently at each operator, each device has to be tested for each operator.

This isn't to say that you have to create 200 versions of the same site; typically, you only need to create a version for each device class. But if you plan to work with an operator, it does mean that you have to test on each device you plan to support.

Where these numbers start to hit your bottom line is in quality assurance (QA). Each release will need to be tested thoroughly before the operator allows you to put it into production. This means that getting a product out on multiple operators for multiple devices means a lot of time pushing buttons on many different devices.

Working "Off Deck"

Now there are actually plenty of cases where you might want to use a full adaptation strategy, or least a pseudoadaptation strategy, which is greater than just detecting and routing, but less than completely dynamic. This is normally referred to in the mobile community as being "off deck," which usually means delivering content to multiple mobile devices over the Web and not through an operator.

The most common reasons to use a full adaptation strategy are:

- You want to deliver the best possible experience to a number of Class B or lower devices where other strategies don't cut it.
- You want to support users outside of the United States, where higher-end devices constitute the majority.
- You want to do anything highly transactional, like billing or payments for content or services, for Class B or lower devices.
- You want to do SMS campaigns that terminate with a mobile website.
- You have a media rich experience, such as images, video, or audio, and need it to render properly on several devices.
- You want to support several nonphone mobile devices, like GPS units, e-book readers, portable gaming consoles, and so on.
- You want to support a number of different devices and contexts, either now or in the future.

None of these reasons are unreasonable; in fact, if you want to take mobile seriously and support anything lower than Class A devices, you probably need to take a look at a full adaptation strategy.

Luckily, we have several existing tools to solve this problem as well. Let's briefly look at several of the common tools out there used for full adaptation.

WURFL

WURFL, or the Wireless Universal Resource File (*http://wurfl.sourceforge.net*), is an open source database of device profiles. It is a massive endeavor—the largest and one of the most active open source projects in mobile. WURFL founder Luca Passani defines the project like this:

> WURFL is an XML configuration file which contains information about capabilities and features of many mobile devices.

As new devices are released, the community (often the handset makers themselves) profiles the device. The result is a detailed database of device attributes, from user agent strings to screen size dimensions, supported media types, and other device characteristics.

Usually developers use WURFL to load the device profiles into their server directives or to create their own conditional logic to dynamically load or replace content based on the requesting device. WURFL provides a free API for PHP-, .NET-, and Java-based applications to query the database and return device information (in the form of "capabilities," where a capability represents a particular device property).

For example, we could use WURFL to detect device information in PHP this way:

```php
<?php echo $_SERVER["HTTP_USER_AGENT"];
$requestingDevice = $wurflManager->getDeviceForHttpRequest($_SERVER);
?>

<ul>
<li>ID: <?php echo $requestingDevice->id ?> </li>
<li>Brand Name: <?php echo $requestingDevice->getCapability("brand_name") ?> </li>
<li>Model Name: <?php echo $requestingDevice->getCapability("model_name") ?> </li>
<li>Xhtml Preferred Markup:
<?php echo $requestingDevice->getCapability("preferred_markup") ?> </li>
<li>Resolution Width:
<?php echo $requestingDevice->getCapability("resolution_width") ?> </li>
<li>Resolution Height:
<?php echo $requestingDevice->getCapability("resolution_height") ?> </li>
</ul>
```

WURFL benefits from the feedback of thousands of mobile developers around the globe. Some of these developers are also contributors of device data and can add device information through the public interface available at *http://wurflpro.com*.

In addition to the APIs, the WURFL project offers developers tools such as WALL and WNG adaptation libraries, which we will come back to a bit later in this chapter.

DeviceAtlas

dotMobi's DeviceAtlas is a premium device database service that leverages multiple device databases, including WURFL. DeviceAtlas is meant to be used in one or all of the following ways:

As a test suite
   A full-featured mobile web testing site that enables you to visit a site from a mobile device and run a series of tests

As a web interface to the database

Enables you to search for device properties and interact with the community

As a standards-based API

Enables you to embed in your application an API that is built on W3C recommendations

DeviceAtlas can be used in a variety of ways. For example, DeviceAtlas could be used to detect and redirect devices in PHP this way:

```php
<?php
include 'Mobi/Mtld/DA/Api.php';

$s = microtime(true);

$memcache_enabled = extension_loaded("memcache");
$no_cache = array_key_exists("nocache", $_GET);
if ($memcache_enabled && !$no_cache) {
  $memcache = new Memcache;
  $memcache->connect('localhost', 11211);
  $tree = $memcache->get('tree');
}

if (!is_array($tree)) {
  $tree = Mobi_Mtld_DA_Api::getTreeFromFile("json/Sample.json");
  if ($memcache_enabled && !$no_cache) {
    $memcache->set('tree', $tree, false, 10);
  }
}

if ($memcache_enabled && !$no_cache) {
  $memcache->close();
}

$properties = Mobi_Mtld_DA_Api::getProperties($tree, $_SERVER['HTTP_USER_AGENT']);
//further performance can be gained through caching the properties against the
user-agent as a key (since many requests are likely to come from one device during
its visit)

if($properties[mobileDevice] && !$properties[isBrowser]) {
    Header("Location: /mobi/");
}

?>
```

But DeviceAtlas could also be used as an adaptation tool. In this example, it is used to detect the device, then adapt an image to be optimized for the requesting mobile device:

```php
<?php
include('imageAdaptation.php');
$i='imgc/eye.jpg'; // This is the main image that you want to see in your mobile
screen
$imgurl=convertImage($i); // $imgurl will hold the path of the converted image that
is suitable for your mobile
?>
```

```
<!DOCTYPE html PUBLIC "-//WAPFORUM//DTD XHTML Mobile 1.0//EN"
"http://www.wapforum.org/DTD/xhtml-mobile10.dtd">
<html xmlns="http://www.w3.org/1999/xhtml">
<head><title>Test the API</title></head>
<body>
<img src="<?php echo $imgurl;?>" />
</body>
</html>
```

As Andrea Trasatti of dotMobi describes DeviceAtlas:

> DeviceAtlas is a full ecosystem to test devices, store the results, and use them in your
> application, all in one place, all at your fingertips.

DeviceAtlas is available in a tiered plan, depending on how frequently you need to update your data. A free developer's version is also available for testing at *http://devi ceatlas.com*.

Volantis

One of the larger content adaptation vendors, Volantis—like many adaptation solution providers—maintains its own proprietary device database. It has its own testing laboratory, and tests and records all device capabilities to enable it to work with its primary product. Its device detection service is included in all its service plans, but expect the pricing to be out of reach for all but a few. Volantis does, however, provide an open source version of its Java-based software that includes an older version of its database.

From Volantis:

> Volantis Mobility Server™ is a Java-based development and runtime platform, allowing
> web developers to build and run their own mobile Internet applications across over 6,200
> devices.
>
> Volantis Mobility Server is designed to reduce the complexity of managing mobile con-
> tent, so that developers and content owners can more easily create innovative content
> and services and users can use the device of their choice to access it.

For those serious about full adaptation, especially in an enterprise environment, I recommend giving Volantis a look: *http://www.volantis.com/volantis-mobility-server*.

WALL and WNG

WALL (Wireless Abstraction Library) and WNG (WALL Next Generation) are adaptation libraries written by WURFL's Luca Passani. WNG is the more modern tool, building on the success of WALL as a way to abstract content and adapt it on the fly.

An example JSP page written with WNG might look like this:

```
<wng:head>
  <wng:title text="Radio WNG" />
  <wng:meta httpEquiv="Cache-Control" content="no-cache"/>
  <wng:css_style>
```

```
<wng:css selector="body">
  <wng:css_property name="margin" value="0" />
  <wng:css_property name="border" value="0" />
  <wng:css_property name="font-family" value="arial, sans-serif" />
</wng:css>
<wng:css selector="a">
  <wng:css_property name="color" value="#666" />
  <wng:css_property name="font-weight" value="bold" />
  <wng:css_property name="text-decoration" value="none" />
</wng:css>
<wng:css selector=".label">
  <wng:css_property name="color" value="#222" />
  <wng:css_property name="margin-left" value="10px" />
  <wng:css_property name="font-size" value="10px" />
</wng:css>
<wng:css selector=".input">
  <wng:css_property name="color" value="#222" />
  <wng:css_property name="margin-left" value="10px" />
  <wng:css_property name="font-size" value="10px" />
</wng:css>
  </wng:css_style>
</wng:head>
```

Once the device is detected, WNG dynamically replaces the tags with the markup appropriate for the device.

WALL and WNG are open source and available for free for both Java- and PHP-based servers. They are ideal for developers looking to build their own customized adaptation solution. Learn more at *http://wurfl.sourceforge.net/java/tutorial.php* and *http://wurfl .sourceforge.net/wng/tutorial.php*.

Yahoo! Blueprint

Yahoo! Blueprint is another adaptation service, which allows you to host applications on your own server but use the Blueprint XML markup language, loosely based on XForms, to adapt and deliver them to multiple devices, as shown in Figure 13-4.

Blueprint is a free service from Yahoo! and can be a great way to build services on top of a robust and feature-complete platform. Read more about Blueprint at *http://mobile .yahoo.com/devcenter*.

Netbiscuits

Netbiscuits is a web service, like the others, that allows you to create and manage mobile websites and web applications and then detect, adapt, and deliver the experiences to the right devices (Figure 13-5).

It is available in JSP, PHP, and .NET as both a hosted and a custom API in different pricing tiers. Read more about Netbiscuits at *http://www.netbiscuits.com*.

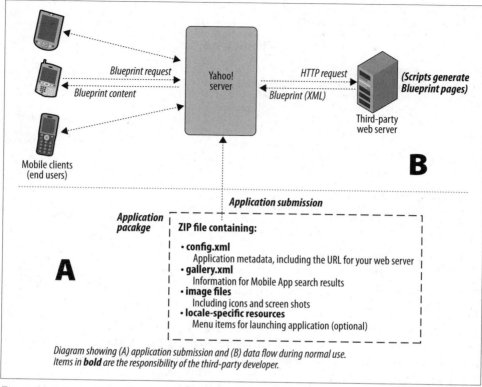

Diagram showing (A) application submission and (B) data flow during normal use.
Items in **bold** are the responsibility of the third-party developer.

Figure 13-4. Using Yahoo! Blueprint to adapt and deliver to multiple devices

MobileAware

MobileAware is an enterprise mobile adaptation system. Like the others, it can detect, adapt, and deliver experiences based on the requesting device (Figure 13-6).

It is available for Java-based servers for a fee. More information is available at *http://www.mobileaware.com*.

Mobify

The folks at Mobify have taken this process of adaptation and made it available for anyone. Using Mobify's web-based tool requires no setup or installation. You can simply point to the content from your desktop website and then apply styling to it. After you point your mobile domain to Mobify, you have an effortlessly adapted site that supports more than 4,000 devices.

The service is offered in tiered monthly pricing, starting with a free limited plan. This is by far the least expensive adaptation solution, and is fairly full-featured compared to the other solutions, which might cost you thousands.

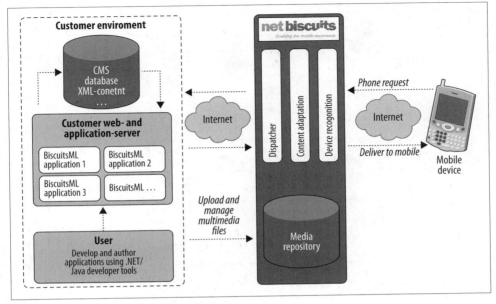

Figure 13-5. The Netbiscuits web service

Find out more about Mobify at *http://www.mobify.me*.

What Domain Do I Use?

Ideally, the user should have to enter only your primary URL—for example, *domain.com*—and your multiserving system should route him to the appropriate experience for his device and context. There are some who believe that the users should control which experience they see, and therefore your mobile experiences should live at a separate URL. I am not one of those people.

I believe routing to the best experience should be the job of technology and give users the option to exit if they so choose. But unfortunately, due to some of the issues with sever-side device routing, as I mentioned before, a separate domain is the only available option. In this case, I offer the following solutions.

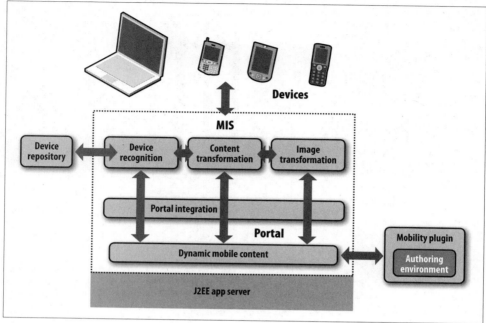

Figure 13-6. The MobileAware web service

m.domain.com

The most common method is to prefix your site domain with a subdomain called "m"—for example, *m.your-domain-name.com*. Now that many sites are providing normal and iPhone versions of their sites, we have *m.domain.com* and *iphone.domain.com*. It's hardly consistent or the best experience for the user, but it is extremely simple to set up.

I recommend at the very least putting all of your mobile content in a single domain, and using one of the previous mentioned detection techniques to route to the best experience. This is preferable to using a URL that the user has to associate with their mobile device.

domain.com/mobile or domain.com/m

Before subdomains were all the rage, most sites used a folder, such as *domain.com/m*. Although it's not as common these days, I think this method works well when combined with subdomains to solve the problem with multiple mobile sites. For example, *m.domain.com* might point to your lowest-common-denominator site and *m.domain.com/iphone* would point to your iPhone site. Any additional versions you create later, such as *m.domain.com/android/*, might simply be added into folders on your *m-domain*. Using a little bit of simple (specifically reverse or *.htaccess*) device detection on the root of the *m-domain* would enable you to redirect your higher-end devices to the appropriate folder.

If you choose not to use folders on your primary domain, I highly recommend setting up *.htaccess* or server redirects on */m* and */mobile* folders to wherever you decide to put your site, just in case.

domain.mobi

The .mobi top-level domain is the ICANN-approved top-level domain specifically for mobile devices and gives publishers the option to use an alternate domain for their mobile site. Instead of using device detection, subdomains, or directors, mobile users go to *domain.mobi*.

If you use a .mobi domain, you can also route all traffic to it, using some simple device detection or server redirects. For example, you could establish server redirects: when a user enters *domain.com/mobile*, she is redirected to *domain.mobi*.

For example, you might consider the following domain mapping:

- *m.domain.com→domain.mobi*
- *domain.com/m→domain.mobi*
- *iphone.domain.com→domain.mobi/iphone*

Taking the Next Step

How do you feel about adapting to mobile devices now? I hope that I was able to report the available options with a minimum of propaganda. Like I said at the beginning: there are no right or wrong answers, only what makes the most sense for your users.

My advice is to start simple with something that can be done in, say, a long weekend. If that means setting up an adaptation system, then go for it. If it means doing something basic, that is OK, too. The most important thing is getting it mobile. Everything else will fall into place after that.

## CHAPTER 14

Making Money in Mobile

The mobile industry is worth an estimated one trillion dollars. Trillion-dollar figures get thrown around quite a bit these days, especially after the worldwide economic crisis of 2008 and 2009. But we forget how extremely large a trillion dollars is. To put it into perspective, a trillion dollars is 1,000×$1,000,000,000 (a billion) or 1,000,000× $1,000,000; certainly, it's not chump change.

Tomi Ahonen, author of *Mobile as 7th of the Mass Media* (futuretext), puts it into perspective this way:

> [The] books business is part of a bigger industry sector called print, which also includes newspapers, magazines, etc. A very big industry indeed. Employing millions on the planet. But is it a trillion dollar industry? No. Only about half that. How about television? Television broadcasts in every country. It's a media giant. Actually, combined with radio, the broadcasting industry is still nowhere near a trillion, only about half that. How about advertising? Surely that is a giant global industry. Yes, it's big, but also advertising is worth roughly half a trillion dollars, in very round terms. Well then the IT industry? No, another half trillion there.

> Let's move beyond the tech and media industries—let's think big. How about bottled water? Drinks, milk, beverages. Let's make this big. Pepsi, Coca Cola, Red Bull. Tropicana orange juice. Heineken, Budweiser, all the wines from France to Australia. Hard liquor, the vodka martini, shaken not stirred. A good Spyside single-malt whisky like a Glenlivet or Cragganmore or Tomintoul. Coffee. Tea. The worldwide beverages industry. Ok, it's big. Yet, it's not worth a trillion dollars in size.

> The world total GDP—the total Gross Domestic Product for the planet—is about 55 trillion dollars. Our economy does not have room for more than a handful of trillion-dollar industries. Look at the countries. Russia, Spain, Canada, Brazil, India, Switzerland—most big countries on the planet—their total national output per year— do not hit one trillion dollars. Italy, France, and Britain are countries whose total domestic output exceeds a trillion dollars. That is the size we're looking at. All of Japan is 4.5 trillion dollars. All of the USA is 14 trillion dollars.

But with all this money flowing into mobile, where does it all go? In Figure 14-1, you can see that voice makes up the majority of the market, taking in nearly $600 billion a year. Hardware is a distant second, with $200 billion a year, messaging $130 billion,

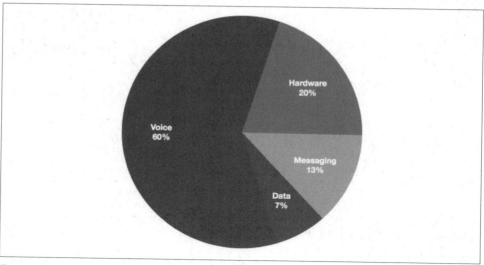

Figure 14-1. Where the money goes in the mobile industry

and data—which accounts for basically everything I've discussed in this book—takes in about $70 billion a year.

Although the data market is certainly a much smaller piece of the pie, it isn't chump change, either. So the question you might be asking is "How can I get in on the action?" How can you make a mobile product and make money from it?

Because we are not operators and we cannot charge users directly for data charges they incur, we have different six basic ways to make money from mobile data, that $70-billion-a-year-and-growing market. These are:

- Sell ringtones
- Sell wallpapers
- Sell games
- Sell downloadable native applications
- Sell subscription content
- Add advertising

The problem is that for all of its benefits, the only way you can make money from the mobile web itself is the last option: add advertising. Mobile advertising works, but it requires lots of views in order to be profitable. If you have a brand that can produce those views on mobile devices, then you can probably earn a decent return. For everyone else, it will hardly make the effort worthwhile.

Technically, you could sell subscription mobile web content as well, but it is problematic and often impractical for all but a few companies. Entering payment information on devices is difficult for the user to do, and the security capabilities on most devices are sketchy at best.

For the only ubiquitous and the most cost-efficient mobile platform, having no means of making money doesn't help justify an investment. It is no wonder that most companies stick to the tried-and-true moneymakers: mobile content such as ringtones, games, and applications.

But what I don't understand is why. Why aren't more people making money from the mobile web? We've learned how to make money from web services and web applications. Users are more than willing to pay the appropriate fees for a useful service, and there is no better medium to provide useful services than mobile. So why haven't we been able to apply this knowledge and experience to the mobile space?

From what I've seen in my journeys in mobile, the obstacles for making money in mobile usually come down to one or more of the following:

- Companies try to support too many devices too quickly. They underestimate the costs of adaptation, testing, and support (a topic I will discuss at length in Chapter 15).
- Downloadable applications are easy one-time wins but become distractions. They can make good money for a short time, but are hardly good long-term investments. Companies divert resources to create a revolving door of downloadable apps to sustain revenue, but end up oversaturating the market instead.
- Companies try to work with operators to put their content in front of a massive number of users. But the operators add too much cost and complexity and give little back in return.
- Companies are unable to produce enough traffic for mobile advertising to earn a profit.

There are numerous benefits to the user in a mobile web strategy, beyond financial gain. Having helpful information and services available to consumers in the mobile context can be used to boost foot traffic to a physical location, increase sales leads, boost brand awareness, and the list goes on. But for the sake of argument, let's say that you want to make a return on your mobile investment—what do you do?

In this chapter, I discuss three approaches to making money in mobile technology—specifically, on the mobile web. However, I should warn you first that there is no perfect formula. In fact, if I had a foolproof answer here, I would most likely be sitting on a beach someplace and not writing a book. How to make money in mobile is an age-old question and one that the entire mobile community has struggled with for over a decade.

Obviously, there are several models that work; otherwise, mobile data wouldn't be that $70-billion piece of the trillion-dollar pie. It is about finding the right model for your business and for your users. And in my experience, it has taken a little luck and good timing.

Working with Operators

As discussed in Chapter 2, the operator is at the center of the entire mobile ecosystem. For better or worse, everyone—even the device makers—have to play by the operator's rules. Every aspect of the industry is guided by its influence. If you are lucky, you will never need to work with an operator at all, but in the right situations, the operator can be an ally and a mutually beneficial friend.

Operators can publish your content to their portal, promote it to their users, and provide a variety of billing and technical services that are often needed. The mobile content trifecta—ringtones, wallpapers, and games—makes up the majority of revenue in this area. For all other content, it can be a struggle to get a new site into the portal. But if you can show how your site can improve their bottom line, it is certainly possible.

Working with operators is easily the most common means of making money in mobile. Let's start by understanding some of the key terms and concepts that come up when talking to an operator.

The Deck

For nonsmartphone devices, it all starts with the *operator portal*, also referred to as the *deck*, which is the default site that is loaded when a mobile web browser is started (see Figure 14-2). It is analogous to the start page or home page that you see when you launch a web browser, but it is provided by your operator. The origin of this term comes from Hypercard and from the later WML development used to create mobile portals. Both languages use the "card" metaphor to denote different pages within a "deck" or site.

The deck is often positioned by the operator as a "walled garden," meaning that you have everything you could possibly need, but you can't leave. On most mobile web browsers, you can, in fact, leave by just entering a URL into your browser, but some operators intentionally hide that function in mobile web browsers provisioned to their network to try to keep the user in their deck.

The deck is the core of the mobile content business; the majority of users browse and purchase mobile content such as ringtones, videos, applications, or games on their device. In addition to their premium content, operators usually include news and information services to attract customers to their deck.

Getting "on deck"

There can be enormous benefits to working "on deck," as it's known, or being linked to from the operator-provided device portal. Trying to develop and promote a mobile website, application, or piece of content is hard work. Although the mobile market itself is vast, building exposure and turning that into a measurable metric of success is quite difficult.

Figure 14-2. The Vodafone deck

Getting your content onto the deck is your basic "stronger together than we are apart" strategy. By pooling resources and providing consumers a storefront of variety that suits their needs and tastes, the deck becomes an easy place to start getting content for your device.

Any website that is listed on the operator deck goes through a rigorous feedback and approval process, most often for downloadable content sites for ringtones, images, or games. Starting with the first site maps and wireframes, the operator reviews your work and provides feedback about how to optimize the experience.

Although it can be extremely frustrating to jump through these hoops, the operator does this for the benefit of your mutual customers. The operator's goal is to reduce problems that lead to support requests, refunds, and lost sales, which is also your goal.

Benefits of the deck

The greatest benefit of the deck is the increased exposure that a promoted central marketplace provides. Remember that nearly every nonsmartphone device that opens a web browser is taken to the operator deck. It is where they start an Internet session. A well-placed item on the operator deck will see millions of views per day. Generating the amount of traffic a deck item receives in a single hour might otherwise take you months or even years to achieve. Most businesses will likely never be able to promote their mobile products on a bigger stage than the operator deck.

Good placement on a deck can equal massive exposure to a product. Top or featured titles, referred to as "deck placement," can see views an order of magnitude higher than the product right beneath it. In other words, if the top application is viewed or downloaded 10,000 times a day, then the application listed 10 down might be downloaded only 100 times a day. Another 10 down and it might have only 10 or fewer views. Top deck placement is determined one of two ways: through sales or through negotiation with the operator. If you are able to negotiate a good deck placement deal, then you may be featured, or temporarily placed at the top of the deck to help build exposure. Think of it like your search engine ranking, but based on sales and not relevance.

For an individual publisher, the notion of publishing content to a marketplace and the high exposure that comes with it can be quite compelling. Given that smaller mobile device screens can usually display only 10–12 deck items per screen, there is always the risk that if you have poor deck placement, you might not be seen at all, but if you have a well-thought-out product with a good design that addresses the mobile context, you have a good chance of earning a return on your investment.

ARPU

Average revenue per user (ARPU, pronounced "are-poo") is the key performance indicator for all things mobile. As voice ARPU becomes more commoditized, operators look to data ARPU for positive quarterly growth needed to impress shareholders. For example, for users on a post-paid, monthly plan, the more content and services the users access, the higher the data ARPU the operator earns.

For many years, data ARPU has been the highest growth area in mobile content subscriptions, where users typically pay a high price per kilobyte downloaded to their device. For example, when a user purchases a ringtone, the operator gets a percentage of the purchase price, and also is able to charge the subscriber for the cost to download the file to her device. Ever wonder why streaming mobile video is so highly touted and heavily marketed by operators? It seems fairly silly that users will want to watch television on their phones. By its nature, it has extremely high data ARPU potential, sending considerable amounts of data to the device that cannot be saved or cached.

Of course, users are wary of having an out-of-control bill. They like to know how much their bills will be every month, or in the case of prepaid, want to make sure that their account credits are prioritized for voice and messaging.

ARPU is the key metric that operators look at in mobile data. Showing you have a plan to boost ARPU typically gets you invited into the club.

BoBo

BoBo means Billing on Behalf of, which allows the operator to bill the consumer on your behalf. With mobile devices, paying for goods or services can be a bit tricky. By the very nature of being mobile, there are common security concerns with transmitting payment information over the air. Although mostly unfounded in newer, more advanced devices, this can certainly be an issue with older devices.

However, the real problem is the awkward task of trying to input credit card information into the device. If you are like me, you might not be able to recall the 16-digit number to enter from memory, so you need to hold up the card and the device and go back and forth between them. Even if you have a better memory than I do, there is the security issue of being out and about where people nearby can look at your mobile screen or which fingers you press.

With BoBo, the user either stores payment information with the vendor, or in the instance of the operator, bills it on your behalf, adds it to your monthly bill, or deducts it from your account balance, in the case of prepaid accounts. Although you lose a percentage of the transaction to the operator, you don't carry the risk of having to manage those transactions, issue refunds, or other accounting hassles. Plus, the consumers have a simple and seamless experience in which they can easily purchase goods and services through their mobile devices.

BoBo is what enables revenue from mobile products. The only other option is to ask the user for credit card information. To use a BoBo strategy, you have to negotiate a deal with each operator you plan to support.

Working with an App Store

Originally, if you wanted to sell something to a mobile user, the only option was to work with the operator, but as smartphones become more popular, we see the emergence of the platform-based app store (see Figure 14-3). The clearest example of this is the Apple App Store, where users can purchase and download games and applications for their iPhones or iPod touches. But we are also seeing app stores like Nokia's Ovi Store; Android Market; and stores for BlackBerry, Palm, and many more devices.

Figure 14-3. The Apple App Store

App stores work essentially the same way as the deck model that operators use. Items must be submitted by the developer and then certified by the app store before they can be placed in the store. Users can then purchase items using the BoBo model, with the app store taking a percentage of each transaction. Then the app store delivers the purchased item to the device.

The top-selling items receive more promotion in the store, and therefore see dramatically more sales. In fact, even though the modern app store experience is delivered in the much bigger screens of smartphones, or even through a desktop app like iTunes, the bulk of the sales goes to the top 10–25 items in the store.

However, there are a few important differences compared to working with operators:

- The app store model requires no negotiation with the app store provider in order to be submitted. If items pass certification, they are entered into the catalog.
- Item placement is purely based on sales, not on negotiation with the app store provider.

- The user can see full screenshots and ratings, and users can read comments before they make a purchase.
- Free content is also available to download.

Apart from the previous list, the app store model isn't that different from the traditional model that came before it (the deck). But this model has one significant drawback: it supports only native applications and not mobile web apps. This means that to deploy your product to each app store, you need to port your application to multiple platforms.

What About the Mobile Web?

Surely we can build a mobile web app and deploy and sell it through an app store? Yes, you can. In Chapter 12, I mentioned the PhoneGap project, which enables you to take a mobile web app and deploy it as a native app for iPhone, Android, and BlackBerry. As it is an open source project, there are also active ports for Nokia and Windows Mobile platforms.

This is possible because many of the smartphone platforms enable you to embed web content into a native application container. PhoneGap provides a quick start, but how to do this from scratch is well documented by each of the major smartphone platforms.

This means you can create a mobile web app that supports multiple mobile devices and distribute and sell it through each of the major app stores. Although it isn't the perfect solution—ideally, you could charge the user in a more direct fashion and not have to go through the app stores at all—but it works.

Add Advertising

The third strategy is to add advertising to your content. Using this model on the mobile web, you are fully in control with no outside dependencies: no negotiation with operators, no app store certification, and no application containers. As mentioned before, the revenue earned from mobile advertising completely depends on the amount of traffic you are able to get to your mobile site or web app. Mobile advertising can effectively work in any mobile product, not just a mobile web app. Many free iPhone apps, for example, use mobile advertising to monetize their use.

The common concern is the impact that advertising can have on the user experience. With such limited and therefore precious screen real estate, placing an ad on the screen can be obtrusive to the user. Good use of mobile ads means placing the appropriate number on the right pages or screens. Find a balance between usability and revenue.

To include these into your products, you simply sign up with one of the major mobile advertising platforms and then add the provided code to your product. The platform then renders ads into your content that are properly formatted for your device.

AdMob and Google AdSense

AdMob (*http://www.admob.com*) is a mobile advertising platform that supports PHP and JSP mobile products for mobile web products and provides an SDK specifically for native iPhone applications. The service provides useful reporting and metrics about the performance of your ads.

AdMob supports a number of different-sized ad units, and will dynamically serve the appropriate-size ad for the requesting device.

AdMob also provides a useful and free mobile metric report on a monthly basis at *http://metrics.admob.com*.

Google provides a mobile version of its AdSense advertising product. Like AdMob, AdSense provides code to include in your product and will render the appropriate ad unit to the device.

In 2007, Russell Beattie, a well-known mobile blogger, published some of the advertising revenue he was earning from Mowser, a product he built and later sold to dotMobi. He ran Google's mobile AdSense product and AdMob, and he ran a mobile advertising platform.

Table 14-1 shows some of the numbers he reported for the AdMob ads that were placed on specific pages.

Table 14-1. Russell Beattie's on AdMob

Date	Page impressions	Clicks	Page CTR	Page eCPM	Earnings
10/11/07	41,069	2,572	6.26%	$1.80	$73.18
10/12/07	38,193	2,840	7.44%	$1.98	$75.00
10/13/07	44,536	3,122	7.01%	$1.73	$76.37

At that time, he mentioned that Google AdSense was not receiving the same click-throughs and began to taper off at $30 a day. He reported that the big reason for the traffic he was receiving was due to placing Google AdSense on his site, which increased his search engine placement. His daily page views went from 5,000 a day to 100,000 a day soon after adding Google AdSense to his product.

The Mobile Marketing Association

The Mobile Marketing Association (*http://www.mmaglobal.com*) provides and maintains a number of mobile advertising guidelines, including ad unit sizes as well as code of conduct for mobile marketing, consumer best practices, and other resources for advertisers and publishers. The Mobile Marketing Guidelines are available at *http://mmaglobal.com/mobileadvertising.pdf*.

For more information on mobile advertising, I highly recommend *Mobile Advertising: Supercharge Your Brand in the Exploding Wireless Market* by Chetan Sharma, Joe Herzog, and Victor Melfi (Wiley). This book covers mobile advertising concepts and strategies in depth.

Invent a New Model

Although each of these models have worked for many in the past, I'm not satisfied. I believe there are many opportunities to monetize mobile content—specifically, the mobile web. We just haven't thought of all the ways to do so yet. As you begin to explore the world of mobile, you will undoubtedly bring new ideass for how to make money in mobile.

Share what you learn, run some tests, and publish some numbers. The more shared justification you can provide to companies, the larger and more innovative the entire space will become, meaning that we all benefit.

Mobile is a trillion-dollar industry. It is massive. However, we are still challenged to make a meaningful profit from the work we produce. With the mobile web, if there aren't better ways to monetize the medium, it will continue to fall prey to whatever the app store or operator monetization model of the week might be. This will encourage more device fragmentation and add more cost to develop and support multiple devices for everyone else.

## CHAPTER 15

Supporting Devices

Imagine a restaurant where you could just walk up and order whatever you felt like—no menu, just anything you want. If you felt like eating a salad, the restaurant would make you a salad. If you felt like a steak, it would make you steak just the way you liked it. If you wanted a banana, peanut butter, and potato chip sandwich, it would make it. Or, if you felt like ratatouille niçoise, in which each ingredient is sautéed separately, layered together, then baked to perfection (the proper way to prepare this delicious dish, by the way), it would make that for you, too. If you are the only customer, then this restaurant would certainly be a dream come true, because the chef would have time to focus and prepare your dish to perfection. It would probably be one of the best meals you've ever had in your life.

But, of course, as more customers enter our fictional restaurant and as more custom orders are placed, the kitchen would become a nightmare. The chef, regardless of his training and expertise, would have to deal with so many variables that not only would it become unmanageable, but the quality would start to decrease. It becomes impossible to create so many variations and still maintain a high degree of quality.

This is the case with the task of testing mobile devices, and unfortunately, we are the chefs. Dealing with one device is easy, but as you start adding more and more devices the variables/variations become too great to manage. Maintaining quality and consistency moves further out of reach.

Many developers tell me, "No problem! Let's be great on one and OK on another." A wise strategy, maybe, for the mobile web, but for native applications, OK is often not quite good enough. In order to get your native applications downloaded onto your users' devices, the applications will need to be certified and approved by at least one organization outside of your own—an opaque process regardless of what channel, platform, device, or operator you intend to work with.

Another lazy strategy is to dismiss personally testing devices altogether and assign them to a QA resource or outsource it entirely. Unfortunately, it is not that easy. Good mobile design and development requires that everyone be a part of ensuring that the work they do ends up on devices correctly, making sure that your experiences are usable and valuable to the user. Testing does not simply occur at the end of the project, but throughout the process.

Keep in mind that there is no one foolproof way of testing your work on devices. It takes flexibility and often some creativity to find the right solution for your resources, project, or company. This chapter discusses some of the various techniques and best practices for testing your work on devices and trying to cost-effectively support as many devices as you possibly can.

Having a Device Plan

I've mentioned numerous times in this book how challenging it is to support multiple devices. This is a problem that shouldn't exist, but it does, and it likely always will. One day in the future, mobile phone platforms will be more alike than they currently are, but by that time we will have many more "mobile" devices to contend with.

I hate to break it to you, but the days of being able to simply test your product on two or three different browsers are over. The browsers and modes in which tomorrow's users will access your content will increase. Our only hope for a sane testing cycle will be the Web and consistent support for web standards. But that is tomorrow; let's talk about today.

Sorting out your device support strategy now prepares you for how you will span your content and services across multiple digital landscapes, including phones, desktops, always-on devices, the social web, and more. The list will only grow over time, and if the past is prologue to the future, then each of these new contextual landscapes will appear faster and faster.

Chapter 11 discussed a device plan that used a class-based matrix. Class A devices are defined as the most advanced web-rendering capabilities, while Class F devices offer a limited experience, and a progressive enhancement technique creates experiences that degrade. This approach works great for prioritizing which devices to support based on their support for standards. But there are other factors to consider in the ideal device plan, such as who your users are and what type of device they will likely have.

Deciding What to Support

As discussed in Chapter 5 regarding the development of a mobile strategy, it is best not to assume that you can support everything. I outlined two methods for determining your support strategy at the beginning of a mobile project. The first is to look at your server logs, seeing which devices currently access your site. The second is to use the niche nature of how devices are marketed to consumers. Device makers make devices with specific consumers in mind—if they are cost-conscious, if they are gadget heads looking for the bleeding-edge features, if they are part of the youth market looking for social networking features. In many cases, we can assume that demographics will match up with device classes. In Table 15-1, I show the assumed devices that each market segment will have and what device class they fall into, based on how the most popular devices are targeted to each segment.

Table 15-1. Devices by market segment

Gender/Age	Class A	Class B	Class C	Class D	Class F
Men					
under 18	High	High	Moderate		
18–24	High	High	Moderate		
25–34	High	High	Moderate		
35–44	High	High	Moderate		
45–54	Moderate	High	High	Moderate	Low
55–64		Moderate	High	Moderate	Low
65+			Moderate	High	Moderate
Women					
under 18	High	High	Moderate	Low	
18–24	High	High	Moderate	Low	
25–34	Moderate	High	Moderate		
35–44		Moderate	High	Moderate	Low
45–54		Moderate	High	High	Moderate
55–64			Moderate	High	Moderate
65+				Moderate	Moderate

Example Device Plans

Creating a device plan is as simple as understanding profit and loss, assuming of course that you think understanding profit and loss is simple. Essentially, certain devices will cost more to develop and test for and certain devices will generate more traffic or downloads. Subtract the estimated cost from your estimated revenue and you have your estimated profit. Target the more profitable device class first.

Unfortunately, trying to come up with dollar values for both cost and revenue estimates can be difficult, especially with larger organizations, where resource cost is rarely tracked. In this case, I recommend using a simple numbered scoring system to indicate the possible magnitude. For example, a cost score of 1 is relatively easy to complete and has minimal impact on a single resource. A cost score of 5 might be five times more complex, taking more time or resources. Table 15-2 shows a numbered score system to determine which devices to support.

Table 15-2. An example device plan with a numbered score system

	Estimated cost	Estimated revenue	Predicted profit
Class A devices	1	5	4
Class B devices	2	4	2
Class C devices	3	3	0
Class D devices	4	2	−2
Class F devices	5	1	−4

In this simple example, the Class A device is the easiest to support and Class F is the hardest, whereas the estimated revenue is the inverse, with the Class A providing more of an impact and Class F less of an impact. When we subtract the cost from the revenue, we see a simple score of what our return on investment might be. In this case, the Class A device provides the greatest return on our investment.

But what about some example scenarios that might not be as simple?

A mobile website

Let's say that you want to build a simple mobile website. You want to support as many devices as you can, but you don't want to spend a lot of money to do so, as maybe the return isn't predicted to be too great. A simple scoring system might look like Table 15-3.

Table 15-3. An example mobile website device plan

	Estimated cost	Estimated revenue	Predicted profit
Class A devices	1	4	3
Class B devices	1	3	2
Class C devices	3	2	1
Class D devices	4	2	−2
Class F devices	5	0	−5

In this case, you can see that Class A, B, and C devices can likely help meet your goals, whereas Class D and F devices cost money. I would approach this scenario by definitely targeting Class A and B devices and testing them throughout the design and development process. With Class C devices, I would use progressive enhancement to allow a comparable experience, but I would probably not test it on multiple devices, or prioritize any bugs of Class C or below. Class D and F would officially be unsupported.

Our initial goal was to support as many devices as possible. Using our estimated revenue (which equals 11), we know that by supporting Class A and B devices, we're targeting 64 percent of our addressable market. If we add the pseudosupported Class C to that, we're covering 82 percent of the addressable market.

But the real magic comes from looking at costs. If you total that column, or in other words, if you were to support all classes of devices, the total would be a unit of 14. But if you just focused support on Class A and B devices, that means that for 14 percent of the cost, you can reach 64 percent of the market—a great number to start.

A mobile web app

Let's look at it another way. If you want to build a mobile web app, and use actual dollar amounts to determine the plan, in this scenario you are likely to see a far greater discrepancy between the device classes. Class A has a far great margin than the other classes (Table 15-4).

Table 15-4. An example mobile web app device plan

	Estimated cost	Estimated revenue	Predicted profit
Class A devices	$20,000	$50,000	$30,000
Class B devices	$35,000	$20,000	−$15,000
Class C devices	$45,000	$2,000	−$43,000
Class D devices	$50,000	$0	−$50,000
Class F devices	$60,000	$0	−$60,000

The Class A device is a sure thing. You know you can see a decent profit from the effort—more than covering costs. Class B devices are nearly twice as much to support than Class A and have the potential to produce less revenue. This time, trying to support both Class A and Class B devices would result in a loss, making less money that it would cost to support in the first place.

A mobile commerce portal

In this final example, we'll look at how to create a mobile commerce portal—a much more complex effort that likely needs more support for more devices, as it wouldn't be fair to users to support just the best devices. Though many of the costs associated with this project would be applied across all device classes, like the supply inventory and creating a shopping cart, look at the costs incurred to design exceptions and test on various devices.

Assume that the costs are about the same as the web app discussed previously, but this time the revenue comes from the middle, more widely adopted devices, not just the high end (Table 15-5).

Table 15-5. An example mobile commerce device plan

	Estimated cost	Estimated revenue	Predicted profit
Class A devices	$20,000	$125,000	$105,000
Class B devices	$35,000	$300,000	$265,000

	Estimated cost	Estimated revenue	Predicted profit
Class C devices	$45,000	$200,000	$155,000
Class D devices	$50,000	$95,000	$45,000
Class F devices	$60,000	$15,000	−$45,000

In this scenario, you can see that the Class B and Class C devices produce the greatest amounts of profit. Though more expensive to support than Class A devices, the sheer number of these popular mid-tier devices produces a dramatic change from our previous device plans.

These are just composite device plans meant for comparative purposes, but hopefully you can see the recurring theme at play here. By understanding your costs of supporting key devices or an entire device class and balancing that against your expected return, you can start to see that supporting some devices just makes more sense than others. With a little research on your users and your market, you can have your own plan in place in no time.

Device Testing

So you have your device plan. Now how do you go about getting devices to test your work on? The simple answer would be to just go buy them, but you guessed it: it isn't that easy. Many devices are subsidized through the operator. So buying a device at the advertised price means buying into a two-year contract that goes with it. Because you need only one contract per operator, this means you have to pay full price for each device you plan to support. At $500–$600 per device unsubsidized, the costs of having multiple devices adds up fast: yet another reason to have a well-researched device plan, so you know which devices to purchase.

Access to Devices

Gaining access to multiple devices is a challenge for every mobile design and developer. I recommend that everyone involved has at least one device, indicative of your primary device class, on the desk when working on the project. This will dramatically reduce the number of assumptions that you have to make during the design and development of the product and can ultimately reduce the amount of time spent testing. This of course can be a device shared among everyone in the office, but if you have multiple offices, or outsourced staff, this can be a problem.

There are alternatives to hands-on device testing, which are discussed later in this chapter. But nothing beats having the device in your hand at the moment of creation. Over the course of my career, I've done it all, and my best mobile products have been the ones where I had access to the intended devices. As you know by now, mobile design is about context, and context is not easily duplicated on the desktop. Testing your work outside the walls of your office will affect how you design.

For example, take an application that uses the accelerometer, like being able to shake the device to load a new page like the iPhone application Urban Spoon (Figure 15-1). How do you test an application where you interact with the mobile nature of the device, if you don't have a device to shake?

Figure 15-1. With Urban Spoon you shake to produce a result—something that is hard to test for without an actual device

Guerrilla testing

The guerrilla testing technique I've been known to recommend in mobile web workshops is to go down to the operator store and test your work on the devices in the store. It isn't exactly something you can do during design or development, but you certainly can confirm how it works on several devices that you can hold in your hand all at once. And as you can find operator stores on just about every other block in most urban areas around the world, you are never very far from a "testing center."

Mobile Monday device libraries

Another route I recommend is talking to your peers at your local Mobile Monday chapter. Mobile Monday is a group of mobilers that meet the first Monday of the month in cities around the world. If you live in a city that has a chapter, and it is the first Monday of the month, then you should put this book down and head over to a meeting. It is a great way to connect with your local mobile community. But for our purposes, you should propose to your local chapter that you start pooling together your collective devices. Create a spreadsheet of who has what devices on what network, then offer to share (and track) them amongst the group, like a library, checking out devices for a day, week, or month. This way, instead of one person buying dozens of devices, you can coordinate with each other so that everyone buys one or two devices and shares them among the group.

Estimating the Testing Effort

One of the worst mistakes you can make is underestimating the time and resources required to test your work on multiple devices. The rule of thumb is that device testing can take anywhere between two and four times the development effort, meaning that for every developer, you need at least two to three QA resources. Or if you are flying solo, then for every week of development, you can easily spend two testing and fixing on multiple devices. This of course depends greatly on your device plan and the range of devices you plan to support.

When developing a single mobile product or supporting a single device class, an extended testing effort can simply result in a delayed release. But when you are supporting multiple classes, the testing effort can hold up resources from moving on to the next phase of development, as shown in Figure 15-2.

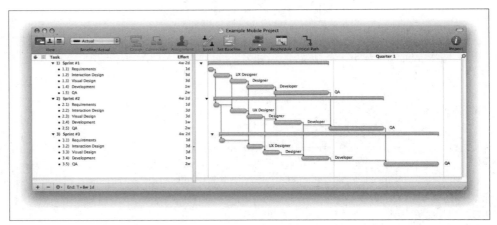

Figure 15-2. An example mobile project plan, showing how lengthy QA cycles can delay phases

This is one reason why I typically recommend focusing on one class at a time, using sprint-like development process. A sprint is a unit of work from the scrum-based methodology that usually lasts anywhere from a week to a month. Support your primary devices in the first sprint, release it, then after a period of time move on to the second class in a second sprint. This frees up your resources and doesn't get them bogged down in development hell during QA. An example project plan is shown in Figure 15-3.

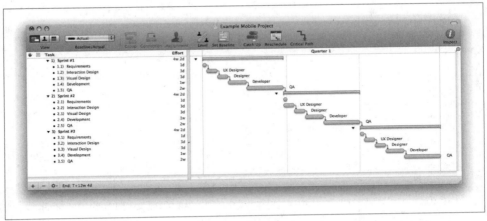

Figure 15-3. An example mobile project plan, separating support into multiple sprints

But if sprints don't work for your project, remember to allow ample time after development is done for testing and device support.

Creating a Test Plan

Creating a test plan for mobile devices means testing for every possible problem that the phone might encounter and that might result in it failing to present the desired content to the user. Because these devices are never from a fixed list like we are accustomed to with the desktop web, that means this can be a pretty big list of things to test. Multiply that by a larger number of devices and you can start to see that this can take some time.

You can get away with lighter testing for mobile web products than native applications, as the latter will need to be certified by a third party in order to get released. So if you don't cover all your bases, your product could fail to pass certification. But in either case, you should create a test plan to make sure that you run your product against a number of routine areas in which the product might fail.

Functional tests

The first step is creating a test plan based on your feature list. Jot down all the features of your application and test them one by one on an actual device, if you have one on hand. Does the application pass or fail? If it fails, then write down what you did to cause it to fail. I encourage people to step away from their computers, go someplace where they can focus, and actually write down their notes on a piece of paper.

You get much better and more detailed feedback when there are fewer distractions. For mobile products, more detailed feedback and finding more bugs earlier on greatly reduces the time and efforts needed to get the product finished and released. I've found that it is easy to simply do a "once-over" that finds only a fraction of the issues you find with an in-depth testing phase.

Context tests

After your features are tested, and you know your application is functionally sound, it is time to move on to testing the mobile context—in other words, testing a few scenarios that can cause your product to fail because it is a mobile device.

Questions to ask about your product during context tests include:

- How does the user experience render on the device? Are there visible issues on the test device? If so, explain them in detail.
- Does it load quickly? Does it load correctly? How many seconds does it take before the user can perform an action? Are you showing a progress indicator if it takes more than a second or two to load? Be sure to test the device at the lowest possible speed. Turn off Wi-Fi, 3G, or any high-speed data connections and load your application using the basic GPRS connection.
- Can you use the physical features of the device as they are intended? Do the soft keys work correctly? Does the view change when you rotate the device orientation? Test any unique physical features that your test device might offer.
- Does it terminate correctly? When you place a phone call to the device, does the application terminate itself correctly? When you resume operation, does it retain your previous state? What about sending a text message? Does it interrupt the session, and can it then be resumed?
- What happens when the device loses its connection? Can users recover their sessions? Test this by starting an action, then go into an elevator or parking garage or someplace where you know you will lose the signal.
- If the device is in offline mode and won't be able to send or receive data, is the user presented with an error or an indication of how to rectify the problem?
- For intense data applications, does the application work when hopping from cell tower to cell tower? Test this by using the application in a car on a highway (with someone else driving, of course).

Creating a Test Portal

A simple but effective trick for mobile web device testing is creating a test portal—basically, a web page with a list of links on it to all your web pages, as shown in Figure 15-4. This provides easy access to all your development servers that you need to test.

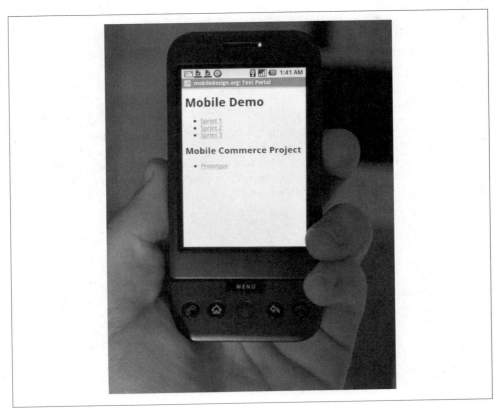

Figure 15-4. An example test portal

Typing URLs into multiple devices is time-consuming, especially development URLs, which can be long. Creating a test portal lets you add URLs for each new product that needs to be tested. You then bookmark that page on each test device, giving you quick access with a fraction of the typing.

Another helpful tool is a URL shortener, like TinyURL (*http://tinyurl.com*) or bitly (*http://bitly.com*), which shorten long URLs to make them mobile-friendly. I actually set up my own service using a tool like Shorty (*http://get-shorty.com*), which you can install and run on your own servers, enabling you to define the short URL. With Shorty, for example, you can make a URL like *http://m.domain.com/testapp/* map to a longer development URL like *http://dev.domain.com/testapp/session?id=12345*.

Desktop Testing

A big advantage of mobile web products is that you can do the majority of your testing from your desktop before ever getting it on a device. You can verify the majority of your markup, styles, and JavaScript, as well doing functional tests on desktop browsers, before putting them on devices and doing your context tests. Desktop testing reduces the time span between developing a feature, testing a feature, and fixing a feature, and ultimately allows you to spend less time dealing with devices.

Frames

Many web browsers have a minimum window size that is larger than your average device screen, making it impractical for testing mobile websites or web apps in a desktop browser. This is where *inline framesets*, or *iframes*, come in handy. Create a web page with an iframe, specifying the dimensions to match your target mobile screen (as shown in the following code), and then add the URL to your mobile project:

```
<iframe src ="mobile/index.html" width="240px" height="320px"
style="border:1px solid;"> </iframe>
```

Because you can embed multiple iframes in a single page, each pointing to the same URL, you can view how your page will look in multiple screen dimensions from a single page. For demonstration purposes, I often use an image of the target device as a background, positioning the iframe where the screen should go, presenting the mobile project as if it were on the device.

Opera

The Opera desktop browser has a Small Screen view (see Figure 15-5) that mimics a mobile screen, loading the handheld media type if available and presenting the page in a narrower format. Because Opera uses the same rendering engine for its desktop browser as the Opera Mobile product, what appears on one is likely to appear the same on the other. However, Opera Mini, the more popular of the two Opera browsers, uses a different rendering engine, but it still comes out close to what you see on the desktop.

Opera's Small Screen view is a great way to see how your work might look on a mobile device—to be taken with a grain of salt. Toggle the Small Screen view by going to View→Small Screen in Opera.

Figure 15-5. Desktop testing using Opera's Small Screen view

WebKit

The WebKit browser engine can also be used for desktop testing for WebKit-based mobile applications. You can download the nightly builds of WebKit (*http://webkit .org*) and run them on your desktop, giving you a close representation of how it may render on the target device, as shown in Figure 15-6.

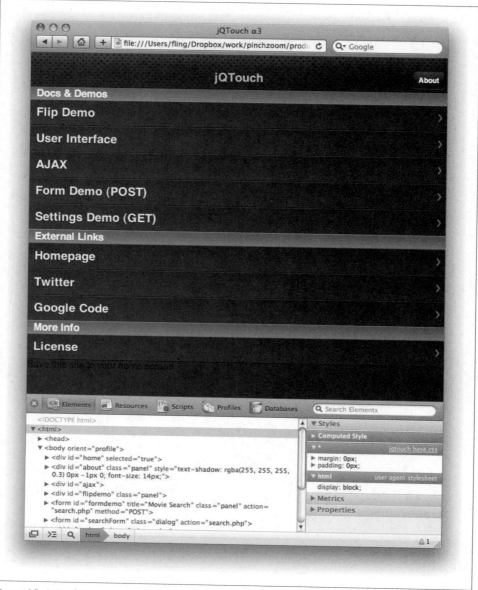

Figure 15-6. Desktop testing with the WebKit browser

Although the version of WebKit for the desktop isn't always the same as the version that's on mobile browsers, it still provides a great means of quickly testing your project prior to loading it onto actual devices.

In addition, WebKit has an excellent debugging tool, called the Web Inspector, which allows you to see how styles are rendered, offline storage information, the page weight, and the estimated time it takes to load resources, all of which are helpful for fine-tuning your mobile web projects, as shown in Figure 15-7.

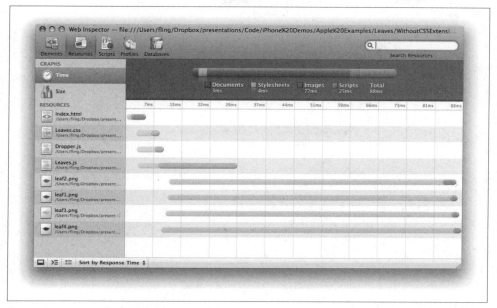

Figure 15-7. Using the Web Inspector to debug mobile web WebKit applications

Firefox

One challenge when using device detection techniques is making sure that they work! The Firefox User Agent Switcher extension (*http://chrispederick.com/work/useragents witcher/*) lets you change the user agent information you send to the server (Figure 15-8). Once you add the data from the supported mobile user agents, you can test how each of your targeted sites renders for the requesting device.

Because few mobile browsers are based on Mozilla's Gecko rendering engine, this form of desktop testing isn't so much for testing how content renders, but is instead for your device detection and content adaptation functional tests.

Also helpful are the Web Developers Toolbar (*http://chrispederick.com/work/webdevel oper/*) and Firebug (*http://www.getfirebug.com*) Firefox extensions to provide XHTML and CSS debugging, similar to the WebKit Web Inspector.

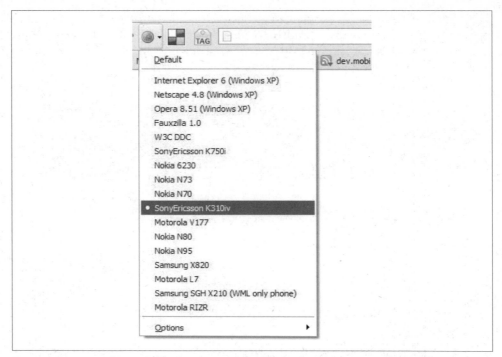

Figure 15-8. Using the Firefox User Agent Switcher to test device detection methods

Collecting User Agents

Test detection and rendering of multiple devices on the desktop means having valid user agents on hand. Given the quantity of devices on the market, there are numerous user agents—sometimes for the same device. The best means of collecting user agents is the open source WURFL database, which contains a large number of user agents submitted by the community and from device makers themselves.

For example, in Chapter 2, I mentioned the numerous mobile browsers found on the Motorola RAZR. Each of those browsers likely has a unique user agent. Searching the WURFL database shows 100 devices and therefore 100 user agents for devices called the "RAZR" (Figure 15-9).

You need only a small fraction of these user agents to test your device detection methods on a desktop browser, but when supporting popular devices like the RAZR, make sure to grab a few variations to make sure all RAZRs get to where they're supposed to go.

Simulators and Emulators

Almost every mobile framework comes with an emulator (Figure 15-10) that allows you to test your work in a virtual environment. Because the hardware in your computer

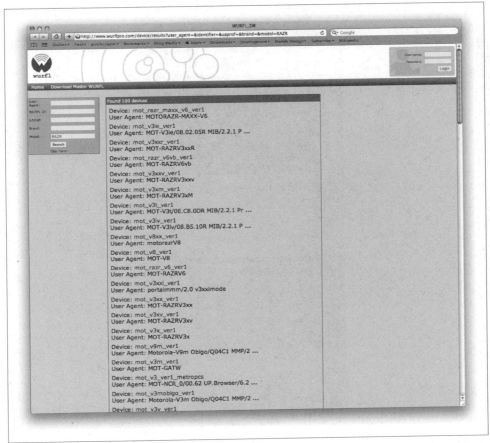

Figure 15-9. The 100 user agents for just the Motorola RAZR

is different than the hardware on the device, it has to run in an emulated environment, and can therefore cause inconsistencies between the emulated environment and the real environment.

Unfortunately, every new device that is released doesn't necessarily get its own emulator. So if we use the RAZR as our example, emulators are only available for a small handful of the total number of RAZRs. In fact, there is no way we can test all available RAZRs using a traditional emulator model.

Furthermore, emulators are often out of sync with the actual devices, sometimes lagging several years behind. For these reasons, I usually recommend that you not rely on emulators for desktop testing, unless they are your only available option. For mobile web products, you can usually get just as accurate and far faster test cycles using desktop browsers as your first-pass and hands-on device testing for your second pass.

Figure 15-10. The dotMobi emulator

Unlike emulators, however, the iPhone Simulator (Figure 15-11) that comes with the free iPhone SDK is an accurate representation of the iPhone environment. Because the iPhone and iPod touch run on Mac OS X, the simulator is able to run the iPhone OS natively on your desktop, not virtually, as long as your desktop is running Mac OS X.

The iPhone Simulator also runs the same version of WebKit that you will find in the devices themselves. Given that the version of WebKit on the device has been customized for the platform, testing in the iPhone Simulator is a far more accurate means of testing mobile web products for the iPhone than simply using the desktop version of WebKit.

Figure 15-11. The iPhone Simulator, which gives you a highly accurate desktop testing solution

Remote Access

Remote access services let you remotely control an actual device through your desktop, but few exist; DeviceAnywhere (*http://www.deviceanywhere.com*) is one. The application taps into the actual device, which you rent through the company's software, and it has most of the devices sold in North America and Europe (Figure 15-12).

Remote access provides a compelling method for testing, with the added convenience of a desktop emulator, in addition to showing the displayed characteristics of the actual device. Remote access doesn't replace the tactile lessons learned from using the device. However, for small publishers with limited resources, it solves the problem of supporting many devices and accessing devices in other countries on their native networks.

Usability Testing

Testing a mobile project with actual users always presents invaluable feedback—it gives you an outside perspective directly from your target users. The more you can test your project, the more data and insight you gain into the potential success or failure of your

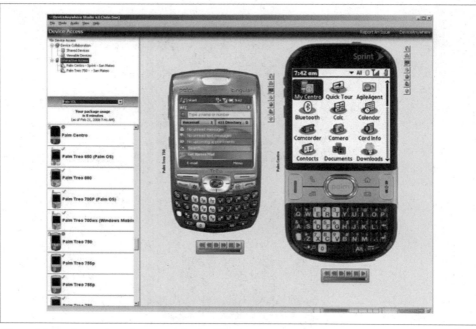

Figure 15-12. DeviceAnywhere

work. As usability guru Jared Spool of User Interface Engineering (*http://www.uie .com*) says, "Talking to two users is better than talking to one."

However, usability testing is not unlike the age-old question "If a tree falls in the forest and there is no one to hear it, does it make a sound?" In psychology this is called reactivity, sometimes referred to as the Observer Effect. When people are observed, their behavior changes, due to the fact that they are being observed. In usability tests, it is common for participants to attempt to please the person conducting the test. Some participants prefer to give overly harsh criticism, trying to find the faults.

The goal of a usability test is to identify actionable faults in the system from subjective behavior and opinion. The researcher must find what data is, or will be, indicative of how the target user will use the system. Unfortunately, this is more art than science.

In mobile development, the challenges of the device can add additional subjectivity. I've found that traditional usability practices fail to address many concerns that the normal user has with mobile technology. Though I'm a big fan of usability testing, especially for mobile products, I find that there are too many variables to produce tangible and actionable results.

The best way to ensure a usable mobile product is to test early, often, and in context—the three things most usability tests skip. The goal is simply to get the experience in front of real users in some sort of analog of the target context. Alternate forms of usability testing, such remote testing, don't usually work well. Café testing—where you get casual unscheduled feedback from participants at a nearby café—works great, given the casual nature of the locale.

Mobile Usability Test Tips and Tricks

Conducting a great mobile usability test comes down to being prepared, keeping context in mind, and being casual and comfortable with the participants. Here are a few tips and tricks that I have found to work well:

Know your users
> Knowing your target users before you start usability testing helps to draw them out while you perform the test. Keep it informal and casual. Don't make them feel like they are being tested. Your posture and attitude during the test will help to define the mood of the room.

Make no assumptions about your participants
> Remember: there are no right or wrong answers in a usability test, and there are no right or wrong participants. Your job is to listen, watch, and learn from your participants. Every participant will tell you something new and valuable. Ask them about how they use their mobile devices and try to integrate this into your findings.

Test early
> The earlier you can test, the better the product. Integrate user feedback into the entire design and development process. Usability testing just at the end of the project does no one any good. Ideally, there should be a user verification and feedback test after each phase of design and development.

Go to the user; don't have them come to you
> Try to perform usability tests in the participants' home, office, or some sort of neutral location. You need to test mobile products in the wild, not in the lab.

Record everything
> It can be a challenge to record screens of mobile devices, but it is important to record as much of the participants' behavior as possible. Small gooseneck digital video cameras can be positioned to capture the device screen as well as the users' reactions. There are a number of different approaches to mobile usability test camera "rigs"; my advice is to play around with the resources available to make them work for you. Just remember that it's more important to listen to your user than it is to fiddle around with technology. Keep it simple!

Have someone else record results
> Whenever possible, have a designated note taker—a third person whose job is to record everything during the test. Her notes and extra perspective will come in handy later when reviewing the test material.

Test often

Usability testing is not a one-time event; it is something you need to integrate into your process. Plan to test as often as you possibly can.

Keep it simple

As I mentioned before, usability testing is more art than science. Keeping your tests simple and casual makes the process easier for you and your participants. It is easy to turn the simple act of talking to your users into an insurmountable chore, full of process and methodology recommended by the "experts." Go easy on yourself and just talk to your users.

Supporting devices, from device plans to usability tests, can certainly be a challenge, but it doesn't have to be. Be prepared, be smart about your plan, and after a few projects, you'll have it down pat.

## CHAPTER 16

The Future of Mobile

I like to think about what's next and what tomorrow's innovations will be. A question I get asked a lot is "What will the future of mobile be?" The best answer I can think of that comes close to capturing the potential that mobile offers is simply "Everything." Tomorrow's innovations will not only involve mobile technology, but they will come from the mobile investments that are made today. This won't be because of the iPhone or Android phones, operators, or the big device makers, but because of people.

We are inherently social beings. We actively seek connections with people every day. The Web is supposed to bring us together, but people feel more isolated than ever. In a survey conducted in 1985, respondents said that they had at least three close friends they felt they could talk to about important issues. In 2006, a Duke University study[*] found that the number of friends people felt they could talk to was down to two people, and 25 percent stated they had no close friends at all.

Mobile technology by its nature is designed to facilitate interaction between people. It is portable, personal, and ubiquitously connected. It enables us not just to communicate in real time, but to collaborate. It is that last point that is so compelling. Phones used to be simply about talking; if we wanted to communicate something, we had to explicitly state it. Collaboration is the iterative back and forth that naturally occurs when two or more people share information.

Whenever I think about how we process information, I come back to the way we learn, for which the technical term is *neurolinguistic programming*, and the three basic sensory modalities that we use to learn: visual; auditory; and kinesthetic, or doing things, like muscle memory. Different people depend on different sensory modalities in order to learn. Some people learn by reading, others by seeing, and some by doing. So therefore in order to teach, you have to appeal to all different types of learners.

Before mobile technology, phones allowed us to collaborate only by hearing voices in an auditory way. Auditory learners make up only about 20 percent of the population; they respond well to information delivered orally and often use phrases like, "Sounds

[*] *http://news.duke.edu/2006/06/socialisolation.html*

good to me" or "That rings a bell." For them, the ability to connect with others verbally is all they need to feel connected and productive with others.

As phones become more like information delivery devices, they start becoming tools for visual learners, which constitute 35 percent of the population. Visual learners use visual stimuli to learn and retain new information. Being able to see concepts illustrated as diagrams or charts and graphs is helpful for this majority learning type. Mobile services like email and web browsers help visual learners communicate with others in an entirely new way.

As touch devices become more prevalent, the third learning type—the kinesthetic learner—is able to incorporate muscle memory and hand-eye movement to process and retain new information. Targets for all three learning types will make the mobile devices of today unique learning and collaboration tools, unlike all other technology before it.

When I think of what the future of mobile can and will be, I see a future in which we present information to all three of these types of learners in multiple contexts. It's a future in which designers and developers can create experiences that start to become predictive, anticipating your next action and transforming the interface to the modalities and location of the user. Devices are smart enough to tackle this today; it is just a matter of putting it to use for companies, products, and users.

However, in order to realize this future platform for all learners, mobile has to overcome many challenges—not just the ones I've discussed in this book, but also the challenges of the Web. Both of these media types are currently on a collision course, fighting each other at times and destined to collide into one another. We need a new type of innovation, which for the wise (and wiley) creates an opportunity for change.

The Opportunity for Change

I see systemic flaws with the Web of today—some deep-seated blight in the industry that makes me want to prognosticate a bleak and bland future. I find it difficult to get excited about "me-too" trends like the next Twitter or Facebook. It's not that these ideas aren't great and that they can't stand some healthy competition, but aren't they just solutions looking for problems? Are they really that important outside a small but vocal cadre that tries to influence others? Are these examples really the ivory pillars of the Information Age?

We saw the birth of the Web 2.0 movement in 2004, but after several years, has it really evolved? How many truly great Web 2.0 sites have emerged since 2005? Three? Four? The mobile industry is even worse. It has been almost 5 years since the mobile bubble, and we are still plagued with the same problems we had 10 years ago. We've had glimmers of hope, but how many game-changing, life-altering, career-defining moments have we had in the mobile web over the past five years? Mobile 2.0, maybe. The iPhone, absolutely.

I sometimes wonder if the mobile web of today is like disco or grunge music: it had a few great songs, some defining moments, produced many one-hit wonders, but in the end it was a vacant trend, filled with so much hype and so heavily capitalized that it fizzled out before anything profound could begin. Technology is not unlike art. It has distinct movements or periods in which a variety of talented people influenced by a shared set of principles defined by the needs and attitudes of the time invent something new, ultimately creating what will become the defining style of the period, an artifact of the collective consciousness of our species.

Despite their similarities, technology has two major defining differences from art. Movements happen a lot faster, making it harder to notice their impact and ultimate outcomes. And it's driven by money: the pursuit of lots of it. Don't get me wrong; money is great, and I'd also like to have lots of it, but the quest for financial gain has a tendency to kill off the passion to create and to innovate. It minimizes the cleansing and necessary creative destruction to identify what's next.

Whoever said money is the root of all evil could have been talking about the Web and mobile technology. I've seen it time and time again, from small garage companies to the big public behemoths—the lust for profit is the cancer of inspired creation. It motivates and encourages bad long-term choices for the sake of short-term gains. The few that have the keen insight to notice these mistakes are rarely rewarded for pointing them out.

Umair Haque, director of the Harvard Havas Media Lab, wrote about the need for a twenty-first-century Industrial Revolution:[†]

> Today's investors, boardrooms, and entrepreneurs are looking for value in all the wrong places. Facebook's game of musical chairs won't solve big economic problems—and neither will making token investments in green tech.

> Where is the next industrial revolution crying out for revolutionaries? Simple: in industries dominated by clear, durable, structural barriers to efficiency and productivity.

I believe that the 2.0 movement is dead. Most disturbingly, I think it died a while ago. Year after year, we keep poking the corpse with a stick, hoping more money will fall out of its pockets. Web 2.0 and Mobile 2.0 by their very names are an iteration of what came before—necessary steps in the evolution of an earlier period. I believe that the time has come for the next movement.

And I'm not talking about "Web 3.0" or anything quite so trite. I believe that Umair Haque is correct. We need a new Industrial Revolution that will define the stepping stones for the next hundred years. We need a deep examination of the impact that the Information Age will have on real people for generations to come.

Mobile is where the conversation starts: it is the introduction to the larger concepts of how to address the user's context in a multidevice environment; how to deal with data

---

[†] *http://discussionleader.hbsp.com/haque/2008/06/a_manifesto_for_the_next_indus_1.html*

portability; what to do about making content accessible to all people, regardless of location, education, or ability; and how to leverage the mobile web, the social web, the desktop web, desktop software, and other emerging technologies to the benefit of your users.

The discussion has started! You have been introduced to mobile, you have the tools to solve problems in the mobile context, and you have a starting point. Now it is up to you to go out and start the twenty-first-century Industrial Revolution. Take mobile technology and integrate it into your business, into your way of thinking, and invent something new and amazing for your users and for yourself.

Index

A

absolute size keywords, text, 190
accelerometers, using in mobile applications, 148
access keys, 180
adjacent sibling selector, 188
AdMob, 274
Adobe Photoshop, 137
advanced attribute selector, 188
advertising, 273
Ahonen, Tomi, 34, 43, 265
Ajax, 157, 226
alt text, 181
American Idol voting, 54
Android, 22
    and WebKit, 202
Android SDK, 24
Appelquist, Daniel, 61
Apple App Store, 113, 119
application context matrix, 88
application framework, 22
applications, 25
attribute selectors, 217

B

background colors and images, 194
Beattie, Russell, 274
Billing on Behalf of (BoBo), 146
Binary Runtime Environment (see BREW)
BlackBerry, 8
blockquote tag, 177
BoBo (Billing on Behalf of), 146
border images, 219
box model sizing, 218

box shadow, 219
BREW (Binary Runtime Environment), 20
    application framework, 23
"brick" style mobile phones, 4

C

cameras in mobile applications, 147
Camino, 201
"Candy Bar Era" of mobile phones, 5
canvas element, 211
carriers, 159
character encoding, 174
character entity references, 179
child selector, 188
cinema, 35
class selector, 188
clearing, 195
Cocoa Touch application framework, 24
context, 46
    C (capitalized), 47
    c (lowercase), 52
context tests, 286
CSS (Cascading Style Sheets), 185–196, 213–225
    animations, 224
    available fonts, 190
    box model, 186
    box properties, 192
    color and backgrounds, 194
    font and text properties, 189
    masks, 222
    positioning and page flow, 194
    selectors, 187
    transforms, 223
    visual effects, 221–225

We'd like to hear your suggestions for improving our indexes. Send email to *index@oreilly.com*.

Wireless CSS and CSS-MP, 186
CSS2, 214
    image replacement, 215
    positioning and page flow, 215
CSS3, 216–221
    attribute selectors, 217
    border images, 219
    box model sizing, 218
    box shadow, 219
    multiple background images, 217
    prefixes, 216
    rounded box corners, 219
    text effects, 220
    text overflow, 220

D

data tables, 183
definition list tags, 178
descendant selector, 187
design patterns, 119
Design4Mobile, 119
device adaptation, 238
device fragmentation, 20, 156
    and ubiquity, 144
device matrix, 170
device plans, 169, 278
    costing device support for mobile web apps,
        281
    devices to support, 278
    examples, 279
    scoring systems, 280
DeviceAnywhere, 295
DeviceAtlas, 256
devices, 18
DHTML, 226
DIAL, 167
Disney and Pixar, 112
display property, 195
Distracted Driver campaign (New Zealand),
        47
div element, 178
doctypes, 173
document structure, 173–176
dotMobi WordPress mobile pack, 252

E

eRuv project, 50

F

Feature Phone Era, 6
Firefox, 201
Firefox User Agent Switcher extension, 291
Fish, Tony, 161
Flash Lite application framework, 23
floats, 195
font replacement techniques, 191
fonts, 190
forms, 183
frames, 183
functional tests, 285

G

games, 79
    native applications and, 147
Google AdSense, 274
GPRS (General Packet Radio Service), 7
graceful degradation, 165
GSM standard, 17
guerrilla testing, 283

H

Haque, Umair, 301
HDML (Handheld Device Markup Language),
        172
headings tags, 176
Herrera, Richard, 227
Hewitt, Joe, 234
Hickson, Ian, 210
high-end versus low-end mobile devices, 32
HTML5, 209–213
    canvas element, 211
    offline data storage, 212
HTML5 and offline applications, 149
Hyatt, Dave, 201, 210

I

IA (see information architecture)
icons, 134
ID selector, 188
iframes, 288
iMac, 44
image maps, 181
images, 180
immersive full-screen application context, 87
information architecture

mobile information architecture (see mobile information architecture)
information architecture (IA), 89–108
informative application context, 84
initiating phone calls, 180
inline framesets, 288
Internet, 13, 36
iPhone, 10, 27
    client-side data storage systems, 212
    creating applications for (see mobile web applications)
    device fragmentation and, 156
    JavaScript and CSS support, 213
iPhone GUI PSD, 233
iPhone simulator, 294
iPhone web applications, 199
    client-side iPhone detection, 208
    fixed footers, compensating for absence of, 227
    substituting clicks and taps for hovering, 225
    WebKit, 200
    XHTML rendering behavior, 208
Iris Browser, 203
iUI, 234

J

Java application framework, 23
Java ME (Java Micro Edition), 20
Java ME (Micro Edition), 77
JavaScript, 157, 196, 225–228
Jobs, Steve, 10
jQTouch, 234

L

largest mobile operators, 15
Lawrence, Jill, 43
layout tables, 183
learning styles, 299
licensed platforms, 20
LiMo, 21
line break tags, 179
links, 179
Linux, 22
locale context, 83
lowest common denominator design strategy, 109

M

Mac OS X, 22
Malkin, Elliot, 50
markup, 172
mass media, 34–40
McGovern, Gerry, 41
media context, 53
Media Queries, 244
MIME types, 174
MMS (Multimedia Messaging Service), 147
.mobi domain, 263
Mobify, 260
mobile
    addressable market, 31–34
    market size and scope, 30
    as a mass medium, 34, 37
    present and future prospects, 29
    target marketing, 33
    unique benefits, 39
Mobile 2.0, 153–157
    creators versus consumers, 160
    interaction needed between mobile and web communities, 159
    JavaScript, 157
    mobile widgets, 158
    reliance on carriers, 159
    rich interactions and battery life, 157
    seven principles of Web 2.0, 154
    web and mobile convergence, 155
    web applications and, 156
    web applications and the user experience, 158
mobile application development
    device adaptation (see multiserving)
mobile application development, principles of, xiii
mobile applications
    context, 81–88
        application context matrix, 88
        immersive full-screen applications, 87
        informative application context, 84
        locale context, 83
        productivity application context, 85
        utility context, 81
    developer control of distribution, 144
    development and consumer expectations, 145
    media matrix, 80
    types, 69–80

games, 79
  medium type, 70
  mobile web applications, 75
  native applications, 77
  SMS, 70
  web widgets, 73
  websites, 71
mobile browser characteristics, Classes A
      through F, 170
mobile business models, 267
  advertising, 273
  app stores, 271
  new models, 275
  operators, 268–271
mobile commerce portals, 281
mobile device
  desktop testing, 288
mobile devices
  by market segment, 279
  context and, 55
  desktop testing
    emulators, 292
    Firefox, 291
    iframes, 288
    Opera Small Screen view, 288
    remote access, 295
    WebKit, 289
  dimensions and PPI, 131
  profusion of platforms, 237
  screen pixels and pixel density, 129
  supported colors, 126
  testing, 277, 282–288
    estimation of test effort, 284
    test plans, 285
    test portals, 287
    user agents, 292
  usability testing, 295
mobile ecosystem, 13
  operator layer, 14
Mobile Fu, 253
mobile industry
  estimated worth, 265
  non-operator revenue streams, 266
  obstacles to financial success, 267
mobile industry, history of, 1
mobile information architecture, 91
  keeping it simple, 93
  site maps, 94
Mobile Marketing Association, 274

Mobile Monday, 154
  device libraries, 284
mobile phones, 18
  evolution, 3–12
  modern mobile phones, 2
Mobile Safari, 202
mobile strategy, 57
  examples, 58
  rules, 59–68
    believe what you see, 60
    constraints aren't first, 61
    context, goals, and needs, 63
    create, 66
    forget what you know, 59
    keep it simple, 67
    not all devices are supportable, 65
    summary, 67
mobile web, 40
mobile web applications, 204
  client-side iPhone detection, 208
  creating, 228–231
  deploying as native applications, 231
  markup, 206–213
  page model, 205
  tools and libraries, 233
mobile web browsers, 155
  WebKit (see WebKit)
mobile web developing, 163
  CSS (see CSS)
  designing for multiple browsers, 165
    DIAL, 167
    progressive enhancement, 165
  designing for multiple displays, 168
    fixed versus fluid displays, 168
    single-column versus multiple column
          layouts, 169
  device plans, 169
    device matrix, 170
  document structure, 173
    auto refresh, 175
    caching, 175
    character encoding, 174
    doctypes, 173
    MIME types, 174
    minimal document structure, 176
    objects and scripts, 175
    page titles, 174
    redirects, 175
    stylesheets, 175

embedded audio and video, 182
external resources, 185
forms, 183
frames, 183
images and objects, 180–182
JavaScript, 196
markup, 172
page sizes, 185
tables, 182–183
validating markup, 184
web standards, 164
XHTML-MP and XHTML Basic (see
        XHTML-MP)
MobileAware, 260
modal context, 54
Moore, Andy, 248, 252
Motorola DynaTAC, 4
Motorola RAZR (V3) phone, 7
Mulmash, Greg, 249
multiserving, 238
    choosing the domain, 261
    device targeting strategy, 247–253
        browser detection scripts, 248
        device detection, 248
        dotMobi WordPress mobile pack, 252
        htaccess-based device detection, 250
        JavaScript-based device detection, 251
        Mobile Fu, 253
        reverse device detection, 251
        Switcher device detection layer, 250
        WordPress mobile plugin, 252
    do nothing strategy, 242–244
        Media Queries, 244
    full adaptation strategy, 253–260
        DeviceAtlas, 256
        Mobify, 260
        MobileAware, 260
        Netbiscuits, 259
        Volantis, 258
        WALL and WNG, 258
        working off deck, 255
        working on deck, 254
        WURFL, 255
        Yahoo! Blueprint, 259
    necessity for, 240
    progressive enhancement strategy, 244–
            247
        handheld media type, 245
        stylesheet layering, 246

multitouch events, 227
MWI (Mobile Web Initiative), 242
Myriad Browser, 203

N

native applications, 77
    advantages, 146–150
        charging for it, 146
        filesystem access, 148
        gaming, 147
        location detection, 147
        offline users, 149
        using accelerometers, 148
        using cameras, 147
    cameras, developing for, 147
    web applications, versus, 143
Netbiscuits, 259
networks (see wireless networks)
neurolinguistic programming, 299
Noki S60, 201
Nokia 9000, 8

O

"One Web" concept, 242–243
operating systems, 22
operators, 14, 159
ordered list tags, 178
overflow, 196

P

page model, 205
page sizes, 185
Palm OS, 22
Palm webOS, 150
paragraph tag, 177
Passani, Luca, 238, 256, 258
Pattern Tap, 119
Pearce, James, 252
Phoenix, 201
PhoneGap, 232
PhoneGap project, 150
PHP mobile browser detection scripts, 248
phrase elements, 177
physical context, 52
Pixar and Disney, 112
pixel density, 130
platform applications, 77
platforms, 20

proprietary and open source platforms, 21
pop-up windows, 184
positioning, 195
printing press, 35
productivity application context, 85
progressive enhancement, 165
    techniques, 166
pseudoselectors, 189

R

radio, 36
recordings, 35
Ribot's Little Spender application, 134
Robinson, Thomas, 227
rounded box corners, 219

S

S60 application framework, 23
Safari, 202
selectors (CSS), 187
    compatibility table, 189
services, 26
Short Message Service (see SMS)
simple attribute selector, 188
site maps, 94
Smartphone Era, 8
smartphone programming languages, 77
SMS (Short Message Service), 6, 70
social media, 160
span element, 178
stacking order, 196
structural elements, 178
Switcher, 250
Symbian OS, 9, 22

T

2.5G networks, 7
tables, 182
target marketing, 33
television, 36
text messaging
    origin of, 6
text styling, 191
Treo smartphones, 8
TTML (Tagged Text Markup Language), 172
type selector, 187

U

ubiquity, 40, 143
    application distribution and the developer,
        144
    and consumer expectations, 145
    device fragmentation and, 144
    services, 41
    and the Web, 144
    web applications and, 145
universal selector, 187
unordered list tags, 177
utility context, 81

V

vector graphics, 182
visual design, 109
    color, 125–128
        color palettes, 128
        psychology, 127
    context, 116
    design elements, 116
    design tools, 137
    designing for the best experience, 115
    graphics, 134
        icons, 134
        photos and images, 136
    interpretation, 111
    layout, 121
        and device support, 122
        fixed versus fluid, 124
        review process, 122
    look and feel, 118–120
    message, 117
    devices, designing for, 138
    screen size, designing for, 139
    tent pole products, 112
    typography, 129–133
        font replacement, 133
        readability, 133
        subpixels and pixel density, 129
        type options, 131
    users' expectations, 109
Volantis, 258

W

-webkit prefix, 216
WALL (Wireless Abstraction Library), 258
WAP (Wireless Application Protocol) 1.0, 172

Web, 25
    advantages of developing for, 144
    systemic flaws of, 300
Web 2.0, seven principles, 154
web applications, 75
    advantages, 150
    Mobile 2.0 and, 156
    native applications, versus, 143
    ubiquity and, 145
web browsers
    box properties compatibility, 193
    designing for multiple browsers, 165
    font and text compatibility, 192
    JavaScript compatibility, 197
    positioning and page flow compatibility,
        196
Web Runtimes (WRTs), 24
web standards, 164, 207
web widgets, 73
WebKit, 199–204
    desktop device testing with, 289
    history, 200
    usage as mobile browser, 201
    usage in iPhone and iPod touch, 202
    WebKit-based browsers, 201
WebKit application framework, 24
webOS, 203
Wikitude, 53
Windows Mobile, 21, 22
    application framework, 24
wireless networks, 17
WML (Wireless Markup Language), 172
WNG (WALL Next Generation), 258
WordPress mobile plugin, 252
WURFL (Wireless Universal Resource File),
    255

X

XHTML, 164, 206
XHTML 2.0, 208
XHTML-MP, 172, 208
    doctypes, 173
    document structure and browser classes,
        173
    text elements, 176
XMLHttpRequest, 157

Y

Yahoo! Blueprint, 259

Z

z-index element, 196

About the Author

Brian Fling owns and runs mobiledesign.org, the largest mobile design and development discussion list on the Web. He's been in both the web and mobile industries for close to a decade as an entrepreneur and consultant. Brian has helped big brands navigate the mobile space, and he's worked with a lot of well-funded mobile companies that have failed miserably. Over the years, Brian has learned that his insight into mobile is quite unique, avoiding hype and describing tried-and-true principles and techniques to building cost-effective mobile experiences.

Brian wrote the *dotMobi Mobile Web Developer's Guide*, the first complete guide to mobile authoring. It was a free guide, and though he doesn't have exact numbers, dotMobi informed him it was downloaded "over 15,000 times in the first few weeks."

His intentions in the mobile space are to advocate and build awareness, not to make money. He believes that the mobile web is primed to change everything we think we know about how people search and gather information. His goal is to foster invention and innovation of the next generation of websites in a medium that is device- and context-aware.

Colophon

The animal on the cover of *Mobile Design and Development* is a twelve-wired bird of paradise (*Seleucidis melanoleucus*). It is found largely throughout New Guinea and the adjacent Salawati Island in Indonesia. The bird's name comes from the 12 thread-like strands that extend from the back of its plumage and bend to cover its behind. The male is black with a yellow belly and yellow feathers along its flanks. The female looks quite different from the male, with its brown plumage on its backside and its black belly. Vegetables, fruit, and anthropods, such as insects, arachnids, and crustaceans, comprise its diet.

The cover image is *Cassell's Natural History*. The cover font is Adobe ITC Garamond. The text font is Linotype Birka; the heading font is Adobe Myriad Condensed; and the code font is LucasFont's TheSansMonoCondensed.

Related Titles from O'Reilly

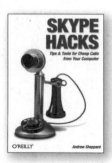

Web Applications

Ambient Findability

Developing Feeds with RSS & Atom

Don't Click on the Blue E!: Switching to Firefox

Dreamweaver 8: The Missing Manual

eBay Hacks, 2nd Edition

eBay: The Missing Manual

Firefox Hacks

Flash 8: The Missing Manual

Google Hacks, *3rd Edition*

Google Pocket Guide

Google Advertising Tools

Google: The Missing Manual, *2nd Edition*

Greasemonkey Hacks

Internet Annoyances

Mapping Hacks

Online Investing Hacks

Podcasting Hacks

Skype Hacks

Talk is Cheap: Switching to Internet Telephones

Using Moodle, *2nd Edition*

Visualizing Data

Web Mapping Illustrated

Web 2.0 Design Patterns

Windows PowerShell: The Definitive Guide

Yahoo! Hacks

Get even more for your money.

Join the O'Reilly Community, and register the O'Reilly books you own.It's free, and you'll get:

- 40% upgrade offer on O'Reilly books
- Membership discounts on books and events
- Free lifetime updates to electronic formats of books
- Multiple ebook formats, DRM FREE
- Participation in the O'Reilly community
- Newsletters
- Account management
- 100% Satisfaction Guarantee

Signing up is easy:

1. **Go to: oreilly.com/go/register**
2. **Create an O'Reilly login.**
3. **Provide your address.**
4. **Register your books.**

Note: English-language books only

To order books online:

oreilly.com/order_new

For questions about products or an order:

orders@oreilly.com

To sign up to get topic-specific email announcements and/or news about upcoming books, conferences, special offers, and new technologies:

elists@oreilly.com

For technical questions about book content:

booktech@oreilly.com

To submit new book proposals to our editors:

proposals@oreilly.com

Many O'Reilly books are available in PDF and several ebook formats. For more information:

oreilly.com/ebooks

O'REILLY®

Spreading the knowledge of innovators

www.oreilly.com

Buy this book and get access to the online edition for 45 days—for free!

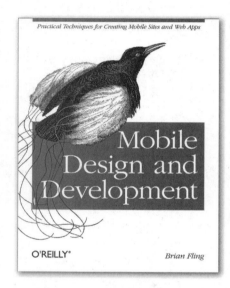

Mobile Design and Development
By Brian Fling
August 2009, $34.99
ISBN 9780596155445

With Safari Books Online, you can:

Access the contents of thousands of technology and business books

- Quickly search over 7000 books and certification guides
- Download whole books or chapters in PDF format, at no extra cost, to print or read on the go
- Copy and paste code
- Save up to 35% on O'Reilly print books
- **New!** Access mobile-friendly books directly from cell phones and mobile devices

Stay up-to-date on emerging topics before the books are published

- Get on-demand access to evolving manuscripts.
- Interact directly with authors of upcoming books

Explore thousands of hours of video on technology and design topics

- Learn from expert video tutorials
- Watch and replay recorded conference sessions

To try out Safari and the online edition of this book FREE for 45 days, go to **www.oreilly.com/go/safarienabled** and enter the coupon code OYTFTZG. To see the complete Safari Library, visit safari.oreilly.com.

Spreading the knowledge of innovators   safari.oreilly.com